Charles Darwin's Theory of Evolution Overthrown

Origin of Species Is Traced - The Cambrian Era
Is Implicated as the Origin of Species

by

Dr. Nyonbeor A. Boley Sr.

The contents of this work, including, but not limited to, the accuracy of events, people, and places depicted; opinions expressed; permission to use previously published materials included; and any advice given or actions advocated are solely the responsibility of the author, who assumes all liability for said work and indemnifies the publisher against any claims stemming from publication of the work.

All Rights Reserved
Copyright © 2017 by Dr. Nyonbeor A. Boley Sr.

No part of this book may be reproduced or transmitted, downloaded, distributed, reverse engineered, or stored in or introduced into any information storage and retrieval system, in any form or by any means, including photocopying and recording, whether electronic or mechanical, now known or hereinafter invented without permission in writing from the publisher.

Dorrance Publishing Co
585 Alpha Drive
Suite 103
Pittsburgh, PA 15238
Visit our website at www.dorrancebookstore.com

ISBN: 978-1-4809-3979-0
eISBN: 978-1-4809-3956-1

This book is dedicated:

To all those who have the desire for truth,
the commitment to find it,
the wisdom to recognize it,
and the courage to accept it.*
Dr. George Cunningham

"Sit down before fact as a little child, be prepared to give up every preconceived notion, follow humbly wherever and to whatever abyss nature leads, or you shall learn nothing."

Thomas Huxley

Table of Contents

Introduction ix

Chapter 1: Science: Mother of Invention 1
Chapter 2: Background: Darwin's Theory of Evolution is Decoded . 5
Chapter 3: Contributors: Evolution is a Flawed Theory 33
Chapter 4: Charles Darwin and Religion: As He Recant 65
Chapter 5: Natural Theology: To the Rescue 77

The Big Bang 95
The Process As I See It 99

Conclusion 113
Important Scientists 125
Glossary 135
Index 139
Footnotes 141
References 143

Introduction

It's about time to remind the thinking world that the controversy between supporters of Darwin's evolutionary theory and creationism is not a new one. However, the current debate between those who accept the theory of evolution as science and those of us who strongly reject it as hypothesis is new. Before our time the debate was between people in favor of the doctrine of divine creation and Charles Darwin's theory of evolution suggesting otherwise that there was no divine creation in his famous work *The Origin of Species* in 1859. Up until that time, the vast majority of Westerners, including almost all scientists, came to accept the biblical view or version of creation. And even after most scientists came to accept Darwin's hypothesis later, large numbers of people still remained opposed to it. In recent time, some of these opponents eventually established the creationist movement that continues to confront, fight, and contradict evolutionary ideas. As far as creationists are concerned, Darwin's "theory of evolution" remains their prime enemy, describing it as "misguided, sinister forces that seemingly appeared out of nowhere to challenge cherished traditional views of humanity and its place in nature." Creationists' opposition to Darwin's theory should rather be based on the fact that it is not a theory but a hypothesis, an hypothesis that blatantly failed to prove the origin of species, and of the human race.

The fact must be stated, however, that if Darwin had not started the battle between evolution and creationism, someone else would have. Contrary to popular belief, he was not even the first thinker or investigator to advance a theory of evolution. Indeed, his version of plant and animal species changing slowly over time did not develop in a vacuum. Instead, his work was in some ways a continuation of a scientific dialogue that had begun many years before. During the two generations preceding Darwin's work, scientists and philosophers had been clearly discussing and debating the age of the Earth, the concept of extinction, the movement of stars, the meaning of fossils, and the doctrine of evolution itself.

Historically, many theories of evolution were advanced during this period, including two by noted French scientists and one by Darwin's own grandfather, Erasmus Darwin.[2] What is more, even these theories were not the first of their kind. Thousands of years earlier, a number of Greek and Roman thinkers had advanced their own hypotheses about the natural progression of living things. These older theorists, however, did not mention anything about a "natural selection and the preservation of favored race" or the "origin of species". Charles Darwin did. But to fully understand and appreciate how the modern controversy over evolution came about, one must examine the development of evolutionary doctrine before, as well as after, the publication of the Origin of Species by Charles Darwin in 1871.

Ancient Greek and Roman Evolutionists

The first attempts to describe the evolution of living things before Darwin occurred in ancient Greece. The sixth-century B.C.E. philosopher-scientist Anaximander was perhaps the first evolutionist (even though as in most cases he may well have gotten his ideas from a predecessor). He suggested that the first living creatures arose in water and that in time these creatures crawled onto the dry land and adapted themselves to their new environments. Moreover, human beings developed the same way. According to this philosopher, "man came into being from an animal other then himself, namely fish, which in earlier times he resembled". This view is no different from Darwin's thoughts as we shall see later in this book.

In the century that followed, another fellow Greek, Empedocles, took Anaximander's views a step further. In Empedocles' thoughts, in dim past, long before human existed, numerous and diverse species of animals had existed. Some of these, he said, were not well adapted for survival in the harsh conditions of their surroundings, so they died out. These are the well documented records on evolution in earlier times but there were some not so well stated by known individuals. Both creative efforts by Anaximander and Empedocles to wonder about how things came into being should be admired. Moreover, our scientific achievement via archeology, paleontology, and genetics today does support some of these discussions by earlier thinkers.

Now, the question may be asked: Why the theory of evolution did not create any serious controversy before Charles Darwin published his masterpiece, *Origin of Species* in 1859? Those earlier evolutionists included Darwin's own grandfather Erasmus Darwin and other Greek scientists-philosophers but they were not credited with advancing evolutionary ideas.

One reason, I believe, that these earlier evolutionary theories did not stir up any significant troubles was the low degree of literacy or interest in matter of science by the average European at that time. All through ancient and medieval times and well into modern times, the majority of people could not read (or did not read well to the point of understanding). Also, very few had access to books. And very few took part in discussions about scientific ideas or were even exposed to such concepts. In ancient times, for instance, there was no mass media, no publishing industry, no bookstores, no internet, no public schools or universal education. Only very rarely did new scientific ideas filter down to the lower classes, whose members made up the vast majority of the population, so average persons were more likely to accept with little apprehension their society's long-held traditional explanation for creation itself, the creation of living things, and other aspects of the natural world. Above all, the Church of England, and the Vatican in particular, did not allow dissent from any member of society at that time on matters of God, creation, and science.

As stated above, religious tolerance out of fear was another powerful reason that evolution did not cause any appreciable controversy in ancient times. One of the chief reasons that Charles Darwin's theory upset people when it first appeared is that they felt such ideas threatened their cherished religious beliefs. In their view, to accept Darwin's ideas was to deny the traditional creation story of Christianity as set down in the Bible.

In the ancient Greco-Roman world, in contrast, few people felt threatened by others' religions or intellectual beliefs. A wide range of faiths and gods proliferated across the Roman Empire; and all were accepted as viable alternative paths to the same internal truths. The ancient Jews and Christians, however, were exceptions. It must be noted that though they were often persecuted, it was not for their religious beliefs but rather for their stubborn exclusivity and insistence that only their own god existed. Their refusal to accept the viability of others' gods and beliefs got them in many troubles. To condemn the religious beliefs of others without giving tangible reasons for your condemnation got the earlier Jews and Christians in hot waters, so to speak. And within these two groups there has been, and there is still, a great divide. Jews don't believe the divinity of Christ. Christians, on the other hand, adjusted their views to believe that Jews are the "chosen people". But this is not the central point in this book. This book is about Charles Darwin's "evolutionary ideas, the origin of species" to be exact.

By Charles Darwin's time, the European world had changed a great deal. Large numbers of people could read. And though most did not read Darwin's book or other scientific works, they read about it in newspapers or heard about in churches, the workplace, or elsewhere. Also, there was very little religious diversity in Darwin's society. Most people were (and remain today) devout Christians; their feelings of religious exclusivity remained strong, and many were suspicious of new ideas that differed from those they had learned as children in Sunday schools and in churches. As you shall see later Darwin himself was a creationist. In such an atmosphere, a quiet discussion among scientists in a private setting was acceptable; but a public dialogue that caused a sudden change in the way people viewed their origins was disturbing and much less acceptable. Darwin added fuel to the fire when he published *The Descent of Man* indicating that the first humans might have evolved from the African Continent. And Darwin was right on the evolution of the human race from Africa but was wrong on the idea of man evolving from primates. But that man evolved from Africa did not sit well with members of Darwin's elite class, as it was unacceptable to many. Why would white Europeans who owned slaves accept the fact that their ancestors were Africans? But to assure his critics, he included a phrase that they were "a naturally selected and superior race." God had preserved a favored race. Still, the added phrase that man might have evolved from Africa, from a much inferior race, did not please everyone. As such, there was a bitter confrontation between the supporters of Charles and those opposed to his "theory" as indicated below.

Confrontation in Oxford

One of the most famous confrontations between Darwin's supporters and those opposed to his theory took place in June, 1860. The event was the annual meeting of the British Association for the Advancement of Science, held at Oxford University. The week-long meeting was attended by many who had read Darwin's book and wanted to discuss his ideas. Darwin himself skipped the meeting, but Joseph Hooker and Thomas Huxley were there. Also in attendance was Richard Owen, who had already expressed his disapproval of Darwin's "theory." An important clergyman, Samuel Wilberforce, the bishop of Oxford, was also present.

At the meeting, Richard Owen spoke out against the idea that humans had descended from apes or other inferior animals. Owen insisted that the differences between the brains of humans and apes made any relationship between them impossible. Huxley replied that the differences were not important! He believed that apes and humans were very likely related and could share the same ancestors. As the meeting progressed, the subject of apes and humans came up in many speeches. Finally, the bishop of the Church of England took the stage.

Bishop Wilberforce was a powerful church leader who also had strong influence in the world of science. Addressing the crowd at the British Association for the Advancement of Science meeting, Wilberforce said that facts did not support Darwin's "theory of evolution." In the bishop's view, humans were distinctly different from other animals, including apes[2].

Humans, as sentience beings who can experience pleasure and pain, endure with a high level creativity could never have evolved from apes. The bishop and Richard Owen were right all along as far as factual evidence goes.

So as you will see in this tiny book, Charles Darwin's theory of evolution in its entirety is on life support, so to speak, because of the wrong conclusion he advanced about the origin of man. On the basis of current research in paleontology and archeology, Darwin's ideas should only be admired and credited for his insight on the evolution of the human race from Africa, but his theory must be overthrown as a non-scientific theory as far as the origin of man is concerned, and *Natural Selection and the Preservation of a Favored Race* should be considered the grandmother of racism in our modern world. And since Charles Darwin was not the first thinker to advance evolutionary ideas, he should not be credited as the sole "father of evolution." He is not. There were many who advanced this thought before him.

This book is divided into two parts. Part One of this tiny book is a refutation of Darwin's theory on the origin of species. Part Two is my feeble attempt to discuss the origin of the human race based on recent discoveries of the Cambrian Explosion and African oral history about the human race.

In writing this book, I have stretched myself far beyond the boundary of my own competence. Even the greatest thinker of all time, which I definitely am not, could not knowledgeably cover such vast ground. But the effort deserves to be made and can't be left just to geneticists, philosophers, science writers, physicists, theologians, or cosmologists. For better or for worse, I have taken the risk, apologizing to the experts for the many instances in which I presumed to trespass on their preserves and even had the temerity to offer my own interpretation on controversial issues including creation.

In Chapter One, the role of science as the "mother of invention" is underscored. Discovery or invention must be based on reason, not revelation as stressed in chapter one.

In Chapter Two, the summary of Charles Darwin's "Theory of Evolution" is given with details, and evidence is provided to discredit it as a non-scientific theory. *Natural Selection and the Preservation of a Favored Race* are also presented as the basis for igniting racism in our civilized world, resulting to the science of Eugenics.

Chapter Three presents list of expert contributors to further discredit Darwin's so-called evolutionary theory, with permission from the publishers.

Chapter Four presents Charles Darwin as a Christian, a Creationist.

Chapter Five introduces "natural theology" as a program to the rescue. This ends Part One of this book.

In Chapter Six my interpretation of creation based on extensive research and study is presented. The book concludes in Chapter Seven, followed by Footnotes and Reference.

Let me state here for the record that had it not been for the encouragement of the editors and publishers this book would never have been published.

Last, but not least, I am grateful to my wife, Nuki Boley, daughter, Joy, and little Al for their patience and encouragement in the writing of this book.

Dr. Nyonbeor A. Boley Sr.

"All venues of inquiry—scientific as well as spiritual—must be pursued in order to arrive at a complete truth concerning the origin of our cosmos, of man, and of all living things."

Dr. Bernard Haisch

Part I

Chapter One
Science: Mother of Invention

At its highest point, the pursuit of science can be profoundly rewarding because the discovery of objective truth is an exhilarating, even mystical, achievement unmatched by any other - one that can truly benefit humanity as perhaps nothing else can, setting itself apart from superstition, witchcraft, and religion.

In order to appreciate the finest value of science, the investigative mind must ask a certain question: What make science, science? Or simply put, what are the essential qualities that define science from all other disciplines?

Prominent among them is an uncompromising humanism and unshakable rationalism. Humanism holds that people have the right to investigate the nature of things to determine their root cause, a right uninhibited by social convention and unrestrained by religious dogma and superstition. Rationalism holds that the means to find that truth is already in humanity's possession, for it is not revelation but reason, an innate power that illuminates the dark corners of humankind's own ignorance, prejudice, and superstition. The inevitable enemy of science is "truth by authority" or authoritarianism - whether human or what some people attribute to be divine revelation - that interposes its arbitrary will between the questioning and investigating mind and the object of its desire.[3]

The force that energizes rationalism is, of course, curiosity - a childlike inclination to ask questions (especially "What," "How," "Why," and "When") and an inquisitiveness that for the scientist must be guided by a mature and resolute refusal to be satisfied by superficial or easy answers, coupled with an abiding determination to persevere patiently in the search for truth. As a result, the life of the scientist may be marked by a heroic individualism that, though at times wounded, is undefeated by disappointment, frustration, and failure. The scientist must also embody a tireless capacity for self-criticism lest he or she mistake pride for certainty. But because the aims of scientific inquiry must often be stated in quantitative terms, the investigator must always be dedicated to precision and exactitude, dispelling wishful thinking and claims that can never be substantiated.

Rejecting the notion of truth by authority, the scientist usually observes the world around him, proposes theories to explain it, and modifies his theories to account for further observations and investigations in the light of new evidence. This is the scientific method, the guiding tool for all scientific inquiry. This is the relevance of this book; once new ideas have proven that Charles Darwin's theory of evolution is, in fact, not a scientific theory and that man did not evolve from molecules, it became timely to present new ideas on the subject matter.

But the road out of the darkness of superstition into the light of reason has not always been an easy one for the scientist. For example, when Vesalius dared to contradict the authority of Galen, he was abused and called a liar and madman; the Montgolfier brothers' claims only met skepticism. Galileo and Copernicus both narrowly avoided following Giordano Bruno to the stake for proposing the heliocentric theory of the solar system, in opposition to accepted Church dogma. Socrates paid with his life for questioning deep-held Church doctrines of hell, heaven, and sin. Yet, the scientific world persevered, and in doing so, lit a beacon for the rest of humanity to follow. This is why it's imperative for scientists of this era to investigate previously-held beliefs that may undermine the work of past and current scientists.

This tiny book you hold in your hand is my attempt to debunk a deep-held belief that Charles Darwin's "evolutionary theory" is a scientific theory and that humans evolved from apes or inferior forms of early life. On the basis of research and study, and benefitting from the advance of molecular biology, I conclude that evolution as proposed by Charles Darwin's *On the Origin of Species by Means of Natural Selection and the Preservation of a Favored Race in the Struggle for Life*, published in 1859 and accepted by some scientists does not meet the standard for a scientific theory. Actual science does not support evolution as a theory so far as the origin of the human race is concerned. Man did not evolve from molecules or apes. *The Descent of Man* published in 1871 by Charles Darwin failed to prove or provide science-based evidence on the origin or descent of man. Scientists of our day must continue the science-based evidence gathering instead of rushing to back-up publications that can't be validated or substantiated. The answer to our origin as a human race may be found in the Cambrian Explosive era from the horn of Africa.

Chapter Two
Background: Darwin's Theory of Evolution is Decoded

This book was inspired by a simple question: Is the theory of evolution as proposed by Charles Darwin a real scientific theory? The answer, as you will find later, is no. The evolutionary theory proposed by Charles Darwin does not meet the criteria of a scientific theory. It is not supported by science, and as far as our understanding of molecular biology is concerned, Darwin's "theory" that man evolved from apes or from a single molecule has no scientific legs to stand on.

One of the most creative men in the twentieth century physics, Albert Einstein, identified two criteria that a scientific theory must meet. The first is that the "theory must not contradict empirical facts." The second criterion concerns the "naturalness" or "logical simplicity" of the premises of the theory. The first criterion, Einstein continues, refers to the "external confirmation" of the theory and the second is concerned with the "inner perfection" of the theory [1].

Agreement between theory and experimental facts is regarded as absolutely essential to all science. So as a result, sometimes the facts precede theoretical explanation. This was the case, for example, with the Brownian motion in physics. Particles suspended in a liquid had been observed zig-zagging through the liquid in a vigorous fashion decades before Einstein provided a quantitative explanation for this phenomenon. At other times, the theory precedes the facts, which was the case when Einstein identified some properties of the photo-electric effects (now quantum mechanics) before experimentalists had established them as facts [1]. In this later example, as is frequently the case, theory identified what facts can be observed. Whatever the sequence, there must be harmony between the theory and experiment. If there is a disagreement and this disagreement persists, experiment becomes, always, the final arbiter and the theory must be brought into agreement with experiment. On the basis of this simple analysis, along with our understanding of molecular biology, Darwin's so-called theory of evolution on the origin of species including humans does not stand a chance. Darwin did not produce the experiment or the science that guided his conclusion of the origin of species and of the human race. There is no logical simplicity or commonsense view of Darwin's thoughts as regard the origin of human beings, particularly. Man did not evolve from apes, a single molecule, amoeba, etc. It is very strange that real thinkers of his day did not challenge the credibility of his theory. Even some well known authors had the temerity to call Charles Darwin a "preeminent scientist" and a "destroyer of myth." What is the basis of Darwin's preeminence in science? But before decoding Darwin's theory in its entirety, a direct quote from Charles Darwin's on the *Origin of Species* is necessary:

Mere chance……might cause one variety (species) to differ in some character from its parents, and the offspring of this variety again to differ from its parents in the same very character and in a greater degree; but this alone would never account for so habitual and large a degree of difference as that species" p-89 (Destroyer of Myths by Andrew Norman).

Essentially, Darwin's theory of evolution seeks to explain the origin of life on Earth as well as the evolution of different species by means of natural selection. Despite the fact that some members of the scientific community have regarded it as a fact for more than a century now, a large number of well learned scientists and others still dispute the evolutionary theory of Darwin as a fact, and as such, various public controversies have resulted from this disagreement. There has never been any well established harmony between the principles of the evolutionary theory and empirical data or even commonsense application concerning the evolution of man from apes. The human body with its complex network of interdependent systems and structures of various cells, tissues, organs, and systems, each with unique function yet all working together to create one smoothly operating entity, could never have evolved from a single molecule to crawl to Earth as such. This is voodoo science, not science proper.

According to evolutionary theory, life began billions of years ago when a group of chemicals, inadvertently or by pure chance, organized themselves into a self-replicating molecule. This tiny molecule gave rise to everything that has ever lived on planet Earth. Different and more complex organisms grew from this simple beginning through mutation of DNA, apparently, evolving from unicellular organisms to more complex multi-cellular organisms, including elephants, whales, and of course, man. Nice try, Charles. But this is scientifically impossible. In the interest of exactitude, let's take a cursory look at the organization of the DNA of humans because such delicate complexity could not have been organized through "pure chance." There is, in my judgment, a "Grand Designer" for this organization.

The DNA that defines every aspect of our bodies is incredibly complex, but in simple and logical terms it can be described as a book of only five letters: *A, G, T, C,* and *U*. Scientifically, to liken DNA to a book, however, is oversimplification of reality or an understatement. The amount of information in the three billion base pairs in the DNA in every human cell is equivalent to that in 1,000 books of encyclopedia size. It would take an experienced typist typing sixty words per minute, eight hours a day, around fifty years to type the entire human genome. And if all the DNA in one human being's trillion cells was put end to end, it would reach the Sun (93,000 thousands square miles away) and back to Earth over 600 times(7). Imagine this complexity evolving from a single molecule, landed on Earth, and gradually evolved into multi-cellular organisms by chance. This cannot be the case.

Aside from the immense volume of information that the DNA contains, consider the likelihood of all the intricate, interrelated parts of this "book" coming together by sheer chance. Critics claim that would be comparable to believing that this publication you are reading happened by accident. Given such scenario, imagine that there was nothing. No Creator of anything whatsoever. Then paper appeared…and ink fell from nowhere onto flat sheets…and shaped itself into perfectly formed letters of the English alphabet. Initially, the letters said something like this: agtcu, uctga, gtauc, tgcauk, gctuak. As you can see, random letters rarely produce words that make sense. But gradually and in time, mindless chance formed them into the order of meaningful words with spaces between them. Then punctuation started mysteriously. Periods, commas, capitals, italics, paragraphs, quotes, and margins also came into being and in correct placements. The sentences then grouped themselves to relate to each other, giving them coherence. Page numbers fell into sequence at the right places, and headers and footnotes appeared from nowhere on the pages, matching the portions of text to which they were related. How possible is this scenario? This scenario reminds me of the "theory of spontaneous generation."

Some molecular biologists, cosmologists, theoretical and particle physicists, and evolutionists do not seem to be embarrassed at all when they say with straight face that the universe - indeed the laws of nature themselves - just appeared for no reason one day, and that human beings in this world evolved from molecules or inferior forms of life. It is necessary perhaps to remember that this was the same case with spontaneous generation - an hypothesis stating that life had arisen from dead matter (as for example, maggots from rotten meat, frogs from mud, flies from bovine manure, fish from the mud of previously dried lakes, and mice from bundles of dirty clothes) until the experiments of Francesco Redi, an Italian scientist, Lazzaro Spallanzani - another Italian scientist, and Louis Pasteur, a French scientist - put to rest the theory of spontaneous generation once and for all.

Then came the idea of the ether. Determining the properties of the ether was a very important issue in the 1905 physics world. The considered judgment of most physicists at that time was that light was a continuous wave. And as a wave, it needed a medium through which it traveled. Since light propagates from place to place at the unearthly speed of 186,000 miles per second (mi/s), the medium through which it traveled had to have the rarest of properties. Against this evidence and judgment, Einstein proposed that light, in fact, is not continuous, light is discontinuous, and that light consisted of particles. Light, when considered as a stream of particles, does not require the ether. Particles do not require a medium. No other physicists, not even his illustrious contemporaries, were thinking about light in this way. After many failed experiments to challenge Einstein's particle theory, the ether story died out. Ether was a myth! Believers of Darwin's evolutionary theory are making the same mistakes again about the origin of species, including human beings. Science is about reason, not revelation, chance, or miracle. The origin of species was unknown to Darwin. We will do well to look at the Cambrian Explosive era for definite answers about our true origin and that of other animals.

There is certainly a "Moving Hand" behind the origin of species and the creation of life on Earth. This "Moving Hand" is the creator of all that exists in our world. It is the universe. We may not be able to describe "it" in words but common sense tells us it exists. Charles Darwin's theory of evolution is simply Darwin's way of expressing his rejection of creationism by the "creator" of the world because of the contradictions he experienced during his five-year voyage around some places. No knowledge should be acquired through fuzzy thinking.

Like Einstein, Denis Diderot reminded us that there are three principal means of acquiring knowledge in the formulation of hypothesis that may result to the proposal of theories: "Observation of nature, Reflection, and Experimentation. Observation collects facts, reflection combines them; experimentation verifies the result of that combination. Therefore, our observation of nature must be diligent, our reflection profound, and our experiment exact." There must be harmony between these three. Charles Darwin failed hopelessly to verify the result of his experiment or conclusion. The advance of molecular biology makes it necessary to discredit Darwin's theory as voodoo science. His observation of life on Earth is cloudy and the result of his experiment contradicts current reality in science and in life. Man did not evolve from apes, unicellular organisms, or molecules.

Charles Darwin's five-year Beagle Voyage around the world allowed him to observe nature and the different species. But how he arrived at the conclusion of the origin of species, including humans, is mind-troubling. Darwin observed nature and reflected on the different species during his five-year trip around most of Western Europe but his conclusion about the origin of species is unscientific and lacks credibility. For example, the science of paleontology was unknown during Darwin's day. Examination of fossil remains, archeological clues or data were not included in Darwin's investigation of the evolution of the human race and other species. This is why after more than 150 years scientists have yet to supply adequate answers to critics that Darwin's theory of evolution is simply bad science. We reason that human beings did not evolve from apes, and that Darwin failed to prove the origin of all living things, including man. The difference between "origination" and "evolution" must be obvious to all scientists.

Darwin's definition of "natural selection" is another cloudy area. We know from biology that natural selection is one biological mechanism that regulates the transmission of whatever forms of information that make it through to the next generation. How did "race" fit into Darwin's selection process? Were certain races eliminated through the natural selection process? Also, we know that during the writing of his "evolutionary theory" Darwin knew nothing about paleontology or archeology, let along genetics, the chemistry and structure of DNA, RNA, the various components of chromosomes, or even the notion of "genes" but at least he knew that organisms, including humans, resemble their parents; that the variation in the appearance of organisms within a single species is inheritable. Why did he include *Natural Selection and the Preservation of a Favored Race*? Whatever Darwin's intention was, scientists in the late 1800s were implementing Darwin's so-called theory of evolution and natural selection as they saw fit. Politicians, economists, and philosophers saw the theory as a convenient explanation for what was happening in Europe and what was to follow in terms of racism and colonization.

According to Darwin's theory, competition was key element in natural selection. Competition also dominated human economic life in the 1800s. Individual businesses were competing fiercely over resources and markets. They used any means available to defeat their rivals. In the United States, industrialists such as Andrew Carnegie and John D. Rockefeller built their business empires by ruthlessly crushing their competitors. Governments gave businesses free rein to pursue their goals. They thought they were naturally selected to destroy anything or anyone who stood in their way!

Before furthering this argument, a direct quotation from Charles Darwin's book *The Origin of Species* is necessary. On page 232 of *The Origin of Species* Darwin wrote:

> Man scans with scrupulous care the character and pedigree of his horses, cattle, and dogs before he matches (breed) them; but when he comes to his own marriage he rarely or never takes such care. Yet he might by selection do something not only for the bodily constitution and frame of his offspring; but for their intellectual and moral qualities. Both sexes ought to refrain from marriage if they in any marked degree one of them is inferior in body (physical appearance) or mind" (from Descent of man, p-138).

He continues:

> Natural selection will produce nothing in one species for the exclusive good or injury of another; though it may be well produce parts, organs, and excretions highly useful or even indispensable, or

highly injurious to another species, but in all cases at same time useful to the owner. Natural selection in each well-stocked country, must act chiefly through competition of the inhabitants one with another, and consequently will produce perfection, or strength in the battle for life, only to the standard of that country. Hence the inhabitants, generally the smaller one, will often yield, as we see they do yield, to the inhabitants of another and generally larger country. For in the larger countries there will have existed more individuals, and more diversified forms, and the competition will have been severer, and thus the standard of perfection will have been rendered higher. National selection will not necessarily produce absolute perfection; nor, as far as we can judge by our limited faculties, can absolute perfection be everywhere found.

Darwin's Supporters Added Race Suppression to the Equation
For many economists and industrial leaders, Darwin's theory of natural selection of a favored race seemed to justify the business practices of the day. Competition and the struggle for existence were essential forces in nature. The same was true in the business world. The strongest, ruthless, and most capable survived and prospered. The weak did not. Any effort or attempt to protect the weak or reduce competition would prevent this natural system from functioning and was therefore resisted.

This social and economic philosophy came to be called Social Darwinism, as Darwin's ideas were used as the reference point. Lamarck's theory that acquired characteristics could be inherited added to the mix. An even more significant influence was the work of the British philosopher and sociologist Herbert Spencer.

Spencer was an early supporter of Darwin's evolutionary theory, especially the ideas of Jean-Baptiste Lamarck. When the origin of species was published in 1859, he accepted Darwin's theory of natural selection of a favored race. Spencer used it in his own writings, along with Lamarckism. In fact, it was Spencer who originated the phrase "survival of the fittest" to further describe natural selection (11). He believed that the same laws that controlled evolution in nature also applied to human society.

"Natural selection and the preservation of a favored race" and" survival of the fittest" ideas were also applied in world politics. During the 1800s, Great Britain and other European powers were expanding their empires. By 1885, leaders from Great Britain, France, Belgium, Portugal, Italy, Germany, and Spain were assembled in Berlin, Germany to partition Africa. Government officials used the idea of natural selection to justify the rule of powerful nations over weaker ones. This argument was partially based on beliefs about differences among racial groups. According to the common view, the native peoples of Africa, Asia, and Australia were members of inferior races. In the eyes of these imperialists, these native peoples were less fit than white Europeans and unable to compete with them. Therefore, European dominance was not only just but natural; they were favored and naturally selected to ruin the lives of so many people. The colonization of these native peoples was a direct result of Darwin's unscientific and misguided idea of *Natural Selection and the Preservation of a Favored Race.*

Summary of the Berlin Conference of 1884-1885 to divide Africa and ruin the African culture and its people:

"The Berlin Conference was Africa's undoing in more ways than one. The colonial powers superimposed their domains on the African continent. By the time actual independence returned to Africa in the 1950s, the realm had acquired a legacy of political fragmentation that could neither be eliminated nor made to operate satisfactorily.

In 1884 at the request of Portugal, Germany Chancellor Otto von Bismark called the major western powers of the world to negotiate questions and end confusion over the control of Africa. Bismark appreciated the opportunity to expand Germany's sphere of influence over Africa and desired to force Germany's rivals to struggle with one another for African territory.

At the end of the conference, eighty percent of Africa remained under traditional and local control. What ultimately resulted was a hodgepodge of geometric boundaries that divided Africa into fifty irregular countries. This new map of the continent was superimposed over the one thousand indigenous cultures and regions of Africa. The new countries lack rhyme or reason and divided coherent groups of people and merged together disparate groups who really did not get along.

Fourteen countries were represented by a plethora of ambassadors when the conference opened in Berlin on November 15, 1884. The countries represented at that time included Austria-

Hungary, Belgium, Denmark, France, Germany, Great Britain, Italy, the Netherlands, Portugal, Russia, Spain, Sweden-Norway (unified from 1814-1905), Turkey, and the United States of America. Of these fourteen countries or nations, France, Germany, Great Britain, and Portugal were major players in the conference, controlling most of colonial Africa at the time.

The initial task of the conference was to agree that the Congo River and Niger River mouths and basins would be considered neutral and open to trade. Despite its neutrality, part of the Congo Basin became a personal and private kingdom for Belgium's King Leopold II and under his rule, over half of the population died because of harsh treatment, I may add.

At the time of the conference, only the coastal areas of Africa were colonized by the European powers. At the Berlin Conference the European colonial powers scrambled to gain control over the interior of the continent. The conference lasted until February 26, 1885 - a three month period where colonial powers haggled over geometric boundaries in the interior of the continent, disregarding the cultural and linguistic boundaries already established by the indigenous African population.

Following the conference, the give and take continued. By 1914, the conference participants had fully divided Africa among themselves into fifty countries under colonial rule.

Major colonial holdings included:

- Great Britain desired a Cape-to-Cairo collection of colonies and almost succeeded through their control of Egypt, Sudan (Anglo-Egyptian Sudan), Uganda, Kenya,

South Africa, and Zimbabwe (Rhodesia), and Botswana. The British also controlled Nigeria, Sierra Leone, and Ghana (Gold Coast).
- France took much of western Africa, from Mauritania to Chad (French West Africa), Gabon, the Ivory Coast, the Republic of Congo (French Equatorial Africa) and Guinea.
- Belgium and King Leopold II controlled the Democratic Republic of Congo (Belgium Congo)
- Portugal took Mozambique in the east and Angola in the west.
- Italy's holdings were Somalia and a portion of Ethiopia.
- Germany took Namibia (German Southwest Africa) and Tanzania (German East Africa).
- Spain claimed the smallest territory - Equatorial Guinea (Rio Muni).

The African Continent has not truly recovered from this racist behavior on the part the European colonial powers. This is why Africa is the only continent without an African-based religion and a political system based an African ideology. This is what the preserved favored race has done to members of the first human race-destroyed their history and identity. The major religions of the Continent of Asia are Budhuism and Hinduism, the Middle East has Judaism, Christianity, and Islam as their major religions. Africa has no religion of its own.

Darwinist imperialists portrayed their colonization of other nations, especially the African Continent as a result of their races being "inferior and "backward." According to such claims, the order of the superior race had to spread across the entire world, and if the world were to progress, the inferior had to be improved. Put another way, the colonialist powers alleged that they were bringing "civilization" to the lands they had conquered. Yet their practices and policies in no way reflected their claims to be well intentioned. Along with their Social Darwinist ideas, the 19th and 20th century colonialist powers brought with them chaos, conflicts, fear and humiliation, rather than well-being, happiness, culture, and civilization. Even if one accepts that the colonialists did provide some benefits for their colonies, still the harm they wreaked was many times greater. Africa is left without its own African religious practices, with forms of government built after their colonial masters!

Courtesy of (http:://geography.about.com/cs/politicalgeog/a/berlinconf.htm)

In the United States, politicians and social philosophers used the same misguided ideas to justify the harsh treatment of Native Americans. These ideas were also called on to support discriminatory laws against former slaves after the Civil War (1861-1865)(12).

The Eugenics Movement - an outgrowth of "natural selection of a favored race"

Other medical officials in the United States had a more drastic solution. Mentally ill people should be sterilized to prevent them from having children. Between 1900 and 1935, at least thirty-five U.S. states passed laws authorizing the sterilization of people in mental asylums. California was one of the leaders in the effort, sterilizing twenty thousand individuals from 1909 to the early 1960s. In total, about sixty-five thousand Americans - including African American women, Native Americans, and common criminals - were subjected to this extreme measure. Most states finally outlawed compulsory sterilization in the mid-1900s. This is the extent to which Darwin's idea of "natural selection" and a "favored race" affected so many people, especially under the banner of Social Darwinism. The eugenics movement in particular caused great suffering to humankind and were all direct outcome of Charles Darwin's misguided evolutionary theory of natural selection.

As stated earlier, natural selection provided a framework for Social Darwinism, but many current ideas contributed to the movement. "Survival of the fittest" became a slogan that represented these varied themes. The ideas of Social Darwinism extended the theory into areas far beyond the world of science. For example, another element of Social Darwinism was the eugenics movement of the early 1900s. The term eugenics was coined in 1883 by Francis Galton, a British scientist and a cousin of Charles Darwin. Galton believed that natural selection should ideally produce human beings who were physically and mentally fit. Galton and other scientists who supported eugenics believed that steps should be taken to keep the human race healthy. They wanted to encourage the "fittest" members of society to have many children. The so-called "unfit" should be discouraged from having sex, to prevent them from reproducing. In the view of most eugenicists, the fittest people were usually well-educated members of the middle-class. The unfit were most often the lower classes, people of color, the poor, and those with mental disabilities.

Charles Darwin's Theory of Evolution Overthrown

In the late 1800s and the early decades of the 1900s, the eugenics movement became widespread in Britain, Europe, and the United States. In 1912 the first International Eugenics Congress was held in London, bringing together many supporters of the movement. Governments took up the cause of eugenics and set up agencies to identify unfit members of society. Those considered unfit were the mentally ill or people with diseases such as epilepsy. In Great Britain the Mental Deficiency Act of 1913 allowed for mentally impaired people to be put into asylums, where they were not allowed to have children. Men and women were kept in separate places to accomplish the goal.

In the United States, the Eugenics Record Office was created in 1910 to identify and locate people suffering from inherited mental illness. Around the same time, Dr. Henry Goddard developed a system of intelligence testing. He set up categories of mental impairment such as "feeble-minded," "moron," and "imbecile." Goddard recommended that people less intelligent be put permanently in institutions. There were no clinical trials conducted and no scientific methods designed to test these hateful acts against humanity. These extreme ideas and movements were a direct outcome of Charles Darwin's unscientific theory of evolution that included natural selection and the preservation of a favored race (Shaking the Foundation- Johnson, S 2013).

These views didn't sit well with science then. People don't get sick because of their race. Diseases are a burden to all humankind, whether yellow, black, or white. This way of thinking has been overthrown by new science-based evidence that people don't get epilepsy or mental illness because of their race and that man did not evolve from apes or single-celled organisms. Though the native peoples might have appeared primitive to Darwin and his believers, man evolved from man, apes evolved from apes, monkeys evolved from monkeys. People are not born inferior. Societies make some people inferior to others. The phrase "natural selection and the preservation of a favored race" as included in Darwin's *Origin of Species* was not based on science. The outcome of *Natural Selection and the Preservation of a Favoured Race in the Struggle for Life* was the Eugenics Movement.

Retrieved with permission from (http://users.adam.com.au/bstett/BEugenics72Bergman73)." History of the Eugenics Movement" starts here:

Though racism can be found throughout history, Darwin was the first to give it an alleged scientific validity. The subtitle of the *Origin of Species*, as stated earlier was the *Preservation of Favoured Races in the Struggle for Life*. Darwin's writings about the "preservation of favoured races," and in particular the unscientific claims in his "Descent of Man" lent support to the Nazis' erroneous belief in the superiority of the Aryan race, and a similar British belief about Anglo-Saxons. In addition, Darwin's theory of natural selection spoke of a fight to the death, "a law of the jungle." Applying it to human societies made conflict and war inevitable between races and nations. A great many prominent figures of the time, from warlike statesmen to philosophers, from politicians to scientists, adopted Darwin's theory. In the twisted road to Auschwitz, Professor Karl A. Schleunes of the North Carolina University history faculty describes how:

"Darwin's notion of the struggle for survival was quickly appropriated by the racists... such struggle legitimized by the latest [so-called] scientific views, justified the racists' conception of superior and inferior peoples…and validated the struggle between them." George Stein, J (2002).

With the claims put forward by Darwin, those who held racist views naturally imagined that they had found a scientific foundation for their views about human classes. But shortly afterwards, science proved that in the same way that Darwin's claims had no scientific validity, a great many movements built around Darwin's ignorant views had committed an enormous error.

With the support it received from Darwinism, the Nazis practiced racism in the most violent manner. Yet Germany was not the only place where so-called "scientific" racism was practiced. A number of racist administrators and intellectuals arose in many countries, particularly in Great Britain and America, racist laws and practices also made a rapid appearance.

Fake evolutionists in the 19th and early 20th centuries held almost totally racist views. Many scientists had no hesitation about openly expressing such opinions. Books and articles written at the time offer the most concrete proof. In *Outcast from Evolution: Scientific Attitudes of Racial Inferiority*, John S. Haller, a professor of history at Southern Illinois University, describes how all 19th century evolutionists falsely believed in the superiority of the white race and that other races were inferior.

Another article in Science magazine made the following comment about some of Haller's claims:

What was new in the Victorian Period was Darwinism…Before 1859, many scientists had questioned whether blacks were of the same species as whites. After 1859, the evolutionary schema raised additional questions, particularly whether or not Afro-Americans could survive competition with their white near-relations. The momentous answer was a resounding NO...The African was inferior because he represented the "missing link" between ape and Teuton. (Sollas, 1879).

William Sollas, a professor of paleontology and geology at Oxford University, set out another of his views in his 1911 book, "Ancient Hunters". He wrote:

> Justice belongs to the strong, and has been meted out to each race according to its strength. It is not priority of occupation, but the power to utilize, which establishes a claim to the land. Hence, it is a duty which every race owes to itself, and to the human family as well, to cultivate by every possible means to its own strength: directly it falls behind in regard it pays to this duty, whether in art of science, in breeding or in organization for self-defense, it occurs a penalty which Natural Selection, the stern but beneficent tyrant of the organic world, will assuredly exact, and that speedily, to the full.

To say that justice belongs to the strong - a grave error will lead to terrible social chaos. No matter what the conditions and circumstances, all people must benefit from true justice, regardless of their color, language, gender, or place of origin. The claims made by Darwinist racists that justice only applies to the strong in no way reflects the truth. Every individual may wish to acquire things of the highest quality and the most attractive for himself and for his society, but he is never justified in ignoring the harm he inflicts on others in doing so. To claim the opposite violates reason and good conscience.

Once the theory of evolution acquired an alleged scientific validity, scientists were able to speak without hesitation of such illusory concepts as "inferior" races and some races being more closely related to apes than to human beings. Despotic dictators such as Hitler recognized such claims as a golden opportunity and killed millions of people because they were "inferior," "inadequate," "flawed," or "sick." One of the main reasons why almost all 19th century evolutionists were racists is that their forerunner, Charles Darwin himself, held such views.

What the Experts Say

Other well known commentators and evolutionary biologists concurred with the conclusion that Charles Darwin theory of evolution does not explain the origin of the human race, and that it is in fact, not a theory or science. For example:

Richard Dawkins says "one thing all real scientists agree upon is the fact of evolution itself. It is a fact that we are cousins of apes and gorillas at the molecular level, and I may add, we did not evolve from apes and gorillas. Evolution is as much a fact as the heat of the Sun. It is not a theory, and for pity's sake, let's stop confusing it as such."

Evolutionary biologist Kirk J. Fitzhugh wrote: Darwin's evolutionary theory cannot be both a theory and a fact. Theories are concepts stating cause-effects relations…One might argue that it is conceivable to speak of evolution as a fact by it being the subject of reference in explanatory hypotheses. In the strictest sense, then, evolution cannot be regarded as a fact even in the context of hypotheses since the causal points of reference continue to be organisms, and no amount of confirming instances for these hypotheses will transform them into facts. While evolution is not a fact, it is also not a single theory, but a set of propositions to a variety of causal questions. An emphasis on associating "evolution" with fact presents the misguided connotation that science seeks certainty. Science does not seek certainty but produces verifiable facts and results.

Michael Denton, in his book *Evolution: A Theory in Crisis* stated that Charles Darwin's theory of evolution is a theory in crisis in the light of tremendous advances we've made in molecular biology, biochemistry, and genetics over the past fifty years. We now know that there are in fact tens of thousands of irreducibly complex systems on the cellular level. Specified complexity pervades the microscopic biological world. Although the tiniest bacterial are incredibly small, weighing less than ten to twelve grams, each is in effect a verifiable micro-miniaturized factory containing thousands of exclusively designed pieces of intricate molecular machinery, made up of one hundred thousand million atoms, far more complicated than any machinery built by man and absolutely without parallel in the non-living world.

We don't need a microscope to observe irreducible complexity. The eye, the ear, and the heart are excellent examples of irreducible complexity, though they were not recognized as such in Darwin's day. Nevertheless, Darwin himself finally confessed, "To suppose that the eye with all its inimitable contrivances for adjusting the focus to different distances, for admitting different amounts of light, and for the correction of spherical and chromatic aberration, could have been formed by natural selection, or evolved from a lifeless single molecule seems, I freely confessed, is absurd in the highest degree."[6]

Darwin's Theory of Evolution is regarded as a slow gradual process. Darwin wrote, "Natural Selection acts only by taking advantage of slight successive variations; she can never take a great and sudden leap, but must advance by short and sure, though slow steps." Thus Darwin conceded that "if it could be demonstrated that any complex organ or system existed, which could not have possibly been formed by numerous, successive, slight modifications, my theory would absolutely break down." Yes, indeed. Darwin's "theory" must be discredited since such a complex system is the human body. Such a complex system would be known as irreducibly complex system. An irreducibly complex system is one composed of multiple parts, all of which are necessary for the system to function. Even if one part is missing, the entire system will fail to function. Every individual part is therefore integral to the whole system. Thus, such a system could not have evolved slowly, piece by piece. Human beings did not evolve from a single molecule after the so-called "Big Bang" and crawl to planet Earth.

Imagine a human being been born without a cardiovascular system, skeletal system, or nervous system but from a single molecule or cell somewhere in the sea and then managed to develop and evolve into a full functioning human being. This is scientifically impossible. Even at the level of embryology, our ultimate height, body shape, eye, skin, hair color, even wrinkles that would someday appear, were all arranged in that forty-six pairs of chromosomes before our fetus status. They did not evolve from climate change. Another fine example is the mousetrap. "The common mousetrap is an everyday non-biological example of irreducible complexity. It is composed of five basic parts; a catch (to hold the bait), a powerful spring, a thin rod called the hammer," a holding bar to secure the hammer in place, and a platform to mount the trap. If any of these parts is missing, or omitted, the mechanism will not work, because each part is integral to the functioning of the mousetrap. The mousetrap is irreducibly complex. Only reducible organ can "grow parts" or evolve overtime.

Still, there are those supporters of Darwin's theory that believe that man is nothing but a bag of chemicals running around. For example, Dr. Werner Gitt quoted Darwin as saying, "the most meaningful result of this book (*Descent of Man*), that man descended from a lowly organized life form, will be a stumbling block for many. I regret that. But there can hardly be any doubts about our descent from savages. According to evolutionary teachings, man's genealogy not only reaches back into the animal kingdom, but right back to simple organic molecules: Primeval soup, primitive slime, primeval cell; singled-celled organisms then became multi-cellular: worms, fishes, amphibians, reptiles, mammals, primitive primates, apes-men, hominids, man."

Nobel prize winner Jacques Monod regards our existence as a necessary consequence of a game of chance [M3, p.137]: "The universe was not pregnant with life nor the biosphere with man.

Our number came up in the Monte Carlo game. Is it surprising that, like the person who has just made a million at the casino, we should feel strange and a little unreal?"

Rupert Riedl also emphasizes the purposelessness of human existence [R2, p. 221] "Man was not planned. In fact, the causal chain of events leading up to man was coincidental. But the results are in the last place necessities...The alternation between necessary chance and accidental necessity has now moved completely inwards: at present the required prejudgment originate inside the central nervous system as preconceived representations. The coincidences of becoming human thus lies in the unpredictability of the convergence of causes. When the first "ugly mammals" originated from earlier reptiles, nobody could have predicted their chances...when the first fishes crawled onto dry land, the question of whether octopus brains were more suitable, was not yet settled." This is a sorry and sad commentary by such a learned individual. Fishes have always lived in water. To date, we still have very primitive and ugly people everywhere in the world including United States, Asia, Australia, and Africa. These individuals did not evolve overtime.

These are some of the views of scientists who based scientific analysis of issues on non-scientific propositions. This kind of discussion or conclusion about the origin of man is unacceptable, as it tends to undermine the credibility of actual science and to trivialize the importance of scientific inquiry. Knowing what is knowable in science, it is very strange that Charles Darwin still has followers in the age of advanced molecular biology, paleontology, and archeology.

The question that inspired the writing of this tiny book, however, has been answered. Charles Darwin's theory of evolution is not based on science. It is not a scientific theory. Man did not evolve from a single inorganic molecule, or from inferior single-celled organisms. Man evolved from man, however primitive his ancestors were. The notion of evolution should and is intimately linked to that of heredity. To be sure, the first evolutionists had only the vaguest ideas on the subject-matter. They not only had no knowledge of DNA, but even the concept of genes was foreign to them. It was only in 1866, seven years after Darwin's publication of the *Origin of Species*, that the founder of genetics, the Austrian monk Gregor Mendel, published the results of his patient observations on the crossbreeding of peas. And it took more than another thirty years before the scientific world rediscovered the now famous "Mendel's Laws" in the obscure journal in which his works appeared. Long before that, however, biologists and even the common run of people obviously knew that children resemble their parents and, in a more general manner, that species perpetuate themselves. Humans give birth to humans, mice to mice, gorillas to gorillas, snails to snails, oak trees to oak trees, apple trees to apple trees, lions to lions, and so on. For evolution to take place completely from outside this arrangement, some element of variation had to be inserted into this continuity, so that eventually man might emerge from an inorganic molecule or from an amoeba. There is no scientific evidence to date to back such "evolution." Respect for truth must take precedence over the regard one may have for the opinions of others.

The foregoing discussions in the next chapter are in support of the conclusion reached in this tiny book by knowledgeable contributors that Darwin's so-called "theory of evolution" has no scientific legs to stand one. We know that Darwin wanted to be a "man of science" but there are no short cuts in science. Science is based on the establishment of a scientific method of a sequence of logical steps. Darwin fell short on these fronts. So, in their own words, the following contributors stated their evidenced-based conclusions.

"One should not pursue goals that are easily achieved. One must develop an instinct for what one can just barely achieve through one's greatest efforts"—

Albert Einstein, 1915.

Chapter 3: Contributors: Evolution Is a Flawed Theory

William A. Dembski

Dr. William A. Dembski, professor of science and theology at Southern Seminary in Louisville, Kentucky is also editor of 'Uncommon Dissent: Intellectuals Who Find Darwinism Unconvincing' and co-editor with Michael Ruse of 'Debating Design: From Darwin to DNA'. At Baylor University, Dembski headed the first intelligent design think-tank. He writes:

> Like alchemists, who hoped to transform lead into gold, evolutionists have failed to identify with sufficient specificity the process by which one organism evolves into another. To believe without proof based on science that life emerged from material causes rather than a designing intelligence is not science, but metaphysics. Without specifying a cause, evolutionists following the foot of Darwin cannot empirically prove that one life emerged from another or that nonliving chemicals combined to form the first life form. As long as evolu-

tion resorts to an unspecified mechanism to explain the origin and evolution of life, it will resemble alchemy more than actual science.

Alchemy explained.

In its heyday alchemy was a comprehensive theory of transmutation describing not only transformations of base into precious metals but also transformations of the soul up and down the chain of being. Alchemy was not just a physics but also metaphysics

Alchemy's metaphysical pretension aside, to include alchemy as part of natural sciences is nowadays regarded as hopelessly misguided. The scientific community rejects alchemy as superstition and wishful thinking and commends itself for having successfully debunked it. For scientists the problem with alchemy is that it fails to specify the processes by which transmutations are supposed to place…

Evolution- alchemy by another name!

Officially, the scientific community rejects alchemy and has rejected it since the rise of modern science by 1650. Unofficially, however, the scientific community has had a much harder time eradicating it. Indeed, I will argue that alchemical thinking pervades the fields of chemical and biological evolution. This is not to deny that biological systems evolve. But unless the process by which one organism evolves into another (or by which nonliving chemicals organize into first living form) is specified, evolution remains an empty word. And giving that such specificity is often lacking, much (though not all) of what currently falls under evolutionary theory is alchemy by another name.

Alchemy followed a certain logic, and it is important to see the fallacy inherent in that logic. The alchemy was not its failure to understand the causal process responsible for a transformation, as should have been the case.

Things transform into other things. Sometimes we can explain the process by which the transformation takes place. At other times we cannot. Sometimes the process requires an intelligent agent, sometimes no agent intelligent agent is required. Thus, a process that arranges randomly strewn Scrabble pieces into meaningful English sentences requires a guiding intelligence. On the other hand, the process by which water crystallizes into ice requires no guiding intelligence - as lowering the temperature sufficiently is all that is needed. It is not alchemy that transforms water into ice. Nor is it alchemy that transforms randomly strewn Scrabble pieces into meaningful English sentences.

What then is the problem with alchemy? Alchemy's problem is its lack of causal specificity. Causal specificity means specifying a cause sufficient to account for the effect in the question. Often we can specify the cause of an effect even if we cannot explain HOW the cause produces the effect.

Alchemy eschews causal specificity. Consider the standard example of alchemical transformation, the transmutation of lead into gold. There is no logical impossibility that prevents potions and furnaces from acting on lead and turning it in to gold. It may just be that we have overlooked some property of lead that, in combination with the right ingredients, allows it to be transformed into gold. But the alchemists of old never specified the precise causal antecedents that would bring about this transformation. Consequently, they lacked any compelling evidence that the transformation was even possible…

Causal specificity was evident in the examples considered earlier: Water cooled below zero degrees Celsius is sufficient to account for turning it to ice. A random collection of Scrabble pieces left in the hands of a literate, non-handicapped English speaker is sufficient to account for Scrabble pieces spelling a coherent English word, phrase, or sentence.. In each of these cases the causal antecedent is specified and accounts for the effect in question. We may not be able to explain how the cause that was specified produces its effect, but we know that it does so nonetheless. This apparently has not been the case with alchemists' views of the transformation of lead to gold.

A Logical Fallacy

Here, then, is the fallacy in alchemy's logic. Alchemy relinquishes causal specificity, yet confidently asserts that an unspecified process will yield a desired transformation. Lacking causal specificity, the alchemist has no empirical grounds for holding that the desired transformation can be effected. Even so, the alchemist remains convinced that the transformation can be effected because prior metaphysical beliefs ensure that some process, though for now unspecified, must effect the desired transformation. In short, metaphysics guarantees the transformation even if the empirical evidence is against it.

Alchemy continues to flourish to this day in the fields of chemical and biological evolution. Whereas classical alchemy was concerned with transforming base into precious metals, evolution is concerned with transforming batches of chemicals into organisms and then organisms into other organisms. I do not want to give the impression that evolution is a completely disreputable concept. The concept has applications that are entirely innocent. Consider, for example, finches evolving stronger beaks to break harder nuts or insects developing insecticide resistance. Evolution in such cases is unproblematic. Why? Because of causal specificity. Micro evolutionary changes like this happen repeatedly and reliably. Given certain organisms placed in certain environments with certain selective pressures, certain predictable changes will result. We may not understand the precise biochemical factors that make such micro evolutionary changes possible, but the causal antecedents that produce micro evolutionary changes are clearly specified. So long as we have causal specificity, evolution is a perfectly legitimate concept.

But what about evolution without causal specificity? Consider, for instance, chemical evolution for the origin of life. For much of the scientific community, the presumption is that life organized itself via undirected chemical pathways and thus apart from any designing intelligence. Yet, unlike causal specificity that obtains for micro evolutionary processes, origin-of-life researchers have yet to specify the chemical pathways that supposedly originated life. Despite vast literature on the origin of life, causally specific proposals for just what those chemical pathways might be are solely absent. This is not to say that there have not been any proposals. In fact, there are too many of them. RNA worlds, clay templates, hydrothermal vents, and numerous other materialistic scenarios have all been proposed to account for the chemical origin of life. Yet, none of these scenarios is detailed enough to be seriously criticized or tested. In short, they all lack causal specificity.

In the absence of causal specificity, the logic of evolution parallels the logic of alchemy. Evolution, like alchemical transformation, is a relational notion. Alchemy never said that gold just magically materializes. Rather, it said that there are antecedents (lead, potions, furnaces) from which it materializes. So, too, evolution does not say that organisms just magically materialize. Rather, it says that there are antecedents (in the case of the origin of life, it posits RNA worlds, clay templates, hydrothermal vents, etc), from which life materializes. Thus, to say that something evolves is to say what it evolves from: just as for the alchemist gold "evolves" from lead plus some other (unspecified) things, so for contemporary origin-of-life researcher organisms "evolve" from suitable (unspecified) batches of prebiotic chemicals.

"X evolve" is therefore an incomplete sentence. It needs to be completed by reading "X evolves from Y." Moreover, the claim that X evolves from Y remains vacuous until one specifies that Y and can demonstrate that Y is sufficient to account for X. Lowering the temperature of water below zero degrees Celsius is causally specific and adequately accounts for the freezing of water. On the other hand, a complete set of the building materials for a house does not suffice to account for a house—additionally what is needed is an architectural plan (drawn by an architect) as well as assembly instructions (executed by a contractor) to implement the plan. Likewise, with the origin of life, it does no good simply to have the building blocks for life (e.g. nucleotide bases or amino acids). The means for organizing those building blocks into a coherent system (e.g. a living organism) need to be specified as well.

Given the pervasive lack of causal specificity in the origin-of-life studies, why are so many origin-of-life researchers supremely confident that material causes are given up to the task of originating life? The singular success of science in elucidating the origin-of-life problem makes this overweening confidence all the more puzzling if we try to understand it [in] light of the skepticism and tentativeness with which the scientific method tells us to approach hypotheses.

Science needs to be free inquiry into all the possibilities that might operate in nature.

On the other hand, if, as I am suggesting, there is a precise parallel between evolution and the alchemy, then this confidence is perfectly understandable, because in that case it flows from a prior metaphysical commitment that is both inviolable and non-negotiable. What prior metaphysical commitment ensures that material causes, though for now unspecified, must effect the desired evolutionary transformations? In the case of alchemy, the prior metaphysical commitment was Neo-Platonism (new followers of Plato). In the case of chemical and biological evolution, the prior metaphysical commitment is, obviously, materialism. Materialism is the view that material causes at base govern the world. Given materialism as a prior metaphysical commitment, it follows that life must evolve through purely material causes. But that commitment, like the alchemists' commitment to Neo-Platonism, is highly problematic, in view of current reality.

Restricting the Range of Solutions

Because the origin of life is an open question, the reference to "purely material causes" lacks, to be sure, causal specificity. But there is the imposition of an arbitrary restriction. The problem with claiming that life has emerged from purely material causes is not that it admits ignorance about an unsolved problem, but that it artificially restricts the range of possible solutions to that problem; namely, it requires that solutions limit themselves to purely material causes. This is an arbitrary and metaphysically driven restriction. Life has emerged via purely material causes. How do we know that? In general, to hypothesize that X results from Y remains pure speculation until the process that brings about X from Y is causally specified. Until then, to impose restrictions on the types of causal factors that may or may not be employed in Y to bring about X is arbitrary and certain to frustrate scientific inquiry.

In this respect evolution is even more culpable than alchemy. Alchemy sought to transform lead into gold, but left open the means by which the transformation could be effected (though in practice alchemists hoped the transformation could be effected through the modest technical means at their disposal). Evolution, on the other hand, seeks to transform nonlife into life and then organisms into very different organisms, but when biased by materialism excludes any place for intelligence or teleology in the transformation exactitude is compromised. Such a restriction is gratuitous given evolution's lack of causal specificity in accounting not only for the origin of life but also for the macro-evolutionary changes supposedly responsible for life's subsequent diversification.

Perhaps materialism will eventually be vindicated and the great open problems of evolution will submit to purely materialistic solutions. But in the absence of causal specificity, there is no reason to let materialism place such restrictions on scientific theorizing. It is restriction like these-typically unspoken, metaphysically motivated, and at odds with free scientific inquiry- that need to be resisted and exposed. Science must not degenerate into applied materialistic philosophy, which is exactly what it does at the hands of today's alchemists- materialistic evolutionists who hold their views not on the basis of empirical evidence but because of a prior metaphysical commitment to materialism. Science needs to be a free inquiry into all the possibilities that might operate in nature. Design, therefore, needs to be kept as a live possibility in scientific discussions of biological origins.

The origin of life is just one instance of evolution without causal specificity. The evolution of human consciousness and language from neurophysiology of primate ancestors is another. The most widely debated instance is the evolution of increasingly complex life forms from simpler ones. Although the Darwinian mutation-selection mechanism is supposed to handle such cases of evolution, it encounters the same failure of causal specificity endemic to alchemy…The lesson of alchemy should be plain: Causal simplicity cannot be redeemed in the coin of metaphysics, be it Neo-platonic or materialistic. And, I may add, to accept Darwin's theory of evolution as authentic is to place scientific inquiry at the mercy of chance itself.

The Theory of Evolution Is Ideology, Not Science

Christoph Schonborn.

Christoph Schonborn, an influential Austrian Roman Catholic Cardinal, participated in the 2005 Papal Conclave that elected Pope Benedict XVI. Schonborn's New York Times editorial excerpted below created considerable controversy among Catholics. It is included here as a contributor to my effort to decode Darwin's "evolutionary theory" as bad science. He writes:

> Any theory that explains the evolution of life as an unguided process left to chance is not science, but ideology. Scientific observation of the evolution of living things reveals a direction and purpose that suppose a Creator. Indeed, human reason drives us to discover our origins and God's existence and His divine purpose. To leave human origins to chance and random variation is to give up on the search to understand the world around us. Such a belief is not science, but the abdication of human intelligence.

Ever since 1996, when Pope John Paul II said that evolution (a term he did not define) was "more than just a hypothesis," defenders of neo-Darwinian dogma have often invoked the supposed acceptance - or at least acquiescence - of the Roman Catholic Church when they defended their theory as somehow compatible with Christian faith, especially the Catholic faith.

But this is not true. The Catholic Church, while leaving to science many details about the history of life on earth, proclaims loudly that by the light of human reason the human intellect can readily and clearly discern and design in the natural world, including the world of living things. This has been the stated position of the Church.

Evolution in the sense of common ancestry might be true, but evolution in the neo-Darwinian sense- unguided, unplanned process of random variation and the natural selection of favored races is not. Any system of thought that denies or seeks to explain away the overwhelming evidence for design in biology is ideology, not science.

The Real Teachings of Pope John Paul II

Consider the real teaching of our beloved John Paul. While his rather vague and unimportant 1996 letter about evolution is alwyays and everywhere cited, we see no one discussing these comments from a 1985 general audience that represents his robust teaching on nature:

> "All the observations concerning the development of life lead to a similar conclusion. The evolution of living beings, of which science seeks to determine the stages and to discern the mechanisms, presents an internal finality which arouses admiration. This finality which directs beings in a direction for which they are not responsible or in charge, obliges one to suppose a Mind which is its inventor, its Creator."

He went on: "To all these indications of the existence of God the Creator, some oppose the power of chance or of the proper mechanisms of matter. To speak of chance for a universe which represents such a complex organization in its elements and such marvelous in its life would be equivalent to given up the search for an explanation of the world as it appears to us. In fact, this would be equivalent to admitting effects without a cause. It would be to abdicate human intelligence, which would thus refuse to think and to seek a solution for its problems," the Pope concluded.

Note that in this quotation the word "finality" is a philosophical term synonymous with final cause, purpose, or design. In comments at another general audience a year later, Pope John Paul concludes, "It is clear that the truth of faith about creation is radically opposed to the theories of materialistic philosophy. These view the cosmos as the result of an evolution of matter reducible to pure chance and necessity."

Finding Design in Nature

Naturally, the authoritative catechism of the Catholic Church agrees with the pope. "Human intelligence is surely already capable of finding a response to the question of origins. The existence of God the Creator can be known with certainty through his works, by the light of human reason." It adds, "We believe that God created the world according to his wisdom. It is not the product of any necessity whatsoever, nor of blind fate or chance."

In an unfortunate new twist on this old controversy, neo-Darwinists recently [as of 2005] have sought to portray our new pope,

Pope Benedict XVI, as a satisfied evolutionist. They have quoted a sentence about common ancestry from a 2004 document of the International Theological Commission, pointed out that Pope Benedict XVI was at the time head of the Commission, and concluded that the Catholic Church has no problem with the notion of "evolution" as used by mainstream biologists—that is, synonymous with neo-Darwinism.

Scientific theories that try to explain away the appearance of design as the result of "chance and necessity" are not scientific at all.

The commission's document, however, reaffirms the perennial teaching of the Catholic Church about the reality of design in nature. Commenting on the widespread abuse of John Paul's 1996 letter on evolution, the Commission cautions that "the letter cannot be read as a blanket approbation of all theories of evolution, including those of neo-Darwinian provenance which explicitly deny top divine providence any truly causal role in the development of life in the universe."

Furthermore, according to the Commission, "An unguided evolutionary process—one that falls outside the bounds of divine providence—simply cannot exist."

Indeed, in the homily at his installation just a few weeks ago [April 24, 2005] Pope Benedict proclaimed, "We are not some casual and meaningless product of evolution. Each of us is the result of a thought of God. Each of us is willed, each of us is loved, and each of us is necessary," the pope concluded.

Throughout history the Church has defended the truths of faith given by Jesus Christ. But in the modern era, the Catholic Church is in the odd position of standing in firm defense of reason as well. In the nineteenth century, the first Vatican Council taught a world newly enthralled by the "death of God" that by the use of reason alone mankind could come to know the reality of the Uncaused Cause, the Mover, the God of the philosophers.

Now at the beginning of the 21st century, faced with scientific claims like neo-Darwinism and the multiuniverse hypothesis in cosmology invented to avoid the overwhelming evidence for purpose and design found in modern science, the Catholic Church will again defend human reason by proclaiming that the immanent design evident in nature is real. Scientific theories that try to explain away the appearance of design as the result of "chance and necessity" are not scientific at all, but as Pope John Paul put it, an abdication of human intelligence.

We must never abdicate human intelligence because we choose to find an easy answer to the aged-old question of the origin of the universe and of our place in it. Logical simplicity, not empirical evidence, is the way out of this dilemma.

Paleontology is primarily concerned with the placement of fossils finds or remains in an evolutionary structure. However, no fossil of intermediate forms have ever been found. At present, there is a full complement of competing hypotheses, but no unified representation exists to date. On informational/theoretical grounds it can be stated that there will never be a phylogenetically based genealogical tree of man because there is no source of new information in evolution. Changing environments (for example, a different climate or changed biotopes) do not qualify as a source of information for new biological structures. In other words, science will never be able to discover or trace the origin of man on its own. Only pure thought can grasp this reality as the ancients dreamed. This is why it is imperative that "all venues of inquiry-scientific as well as spiritual – be pursued together in order to arrive at a complete picture of the truth of the origin of man, of all living things, and of our cosmos." Scientists must not reject religion or spirituality entirely, on the one hand, and spiritualists as well as theologians must not run away from scientific inquiry, on the other hand.

The First Humans: Where Did They Come From?
Dr. Richard Leakey

Dr. Leakey is described by many as the paleontologist whose "fieldwork in East Africa over many decades has contributed so much to our understanding of human evolution." Here he answered the question: What made humans human, and of course contributed to the discussion that Darwin was wrong about the origin of man. He writes:

Anthropologists have long been enthralled by the special qualities of Homo Sapiens, such as language, high technological skills, and the ability to make ethical judgments. But one of the most significant shifts in anthropology in recent years has been the recognition that despite these qualities, our connection with the African apes is extremely close indeed. How did this important intellectual shift come about? Here, I shall discuss how Charles Darwin's ideas about the special nature of the earliest human species influenced anthropologists for more than a century - and how new research has revealed our evolutionary intimacy with African apes and demands our acceptance of a very different of our place in nature other than Darwin's conclusion.

In 1859, in his *Origin of Species*, Darwin carefully avoided extrapolating the implications of evolution to humans. A guarded sentence was added in later editions: "Light will be thrown on the origin of man and his place in history." He elaborated on this short sentence in a subsequent book written eleven years later called *The Descent of Man* published in 1871. Addressing what was still a sensitive subject, he effectively erected two pillars in the theoretical structure of anthropology. The first had to do with where humans first evolved (very few believed him initially, but he was right), and the second concerned the manner of form of that evolution. Darwin's version of the manner of our evolution dominated the science of anthropology up until a few years ago, and it turned out to be very wrong. Man did not evolve from apes.

The cradle of humankind, said Darwin, was Africa. His reasoning was simple:

"In each great region of the world, the living mammals are closely related to the evolved species of the same region. It is, therefore, probable that Africa was formerly inhabited by extinct apes closely allied to the gorilla and chimpanzee: and as these two are man closest allies; it is somewhat more probable that our early progenitors lived on the African continent than elsewhere." Darwin was right. Science proves it. This is where Darwin should get fair credit - to hypothesize that the human race might have evolved from Africa.

We have to remember that when Darwin wrote these words no early human fossils had been found anywhere in Africa as proof for his observation. His conclusion was based entirely on keen intuition. In Darwin's time, the only known human fossils were of Neanderthals, from Europe, and these represent a relatively late stage in human career.

Anthropologists disliked Darwin's suggestion intensely, not least because tropical Africa was regarded with colonial disdain: the Dark Continent was not viewed as a fit place for the origin of so noble a creature as Homo Sapiens. How dared he thought the white man evolved from Africa, let me add. When additional fossils began to be discovered in Europe and Asia at the turn of the century, yet more scorn was heaped on the idea of an African origin. As more and more evidence mounted supporting the idea of our African origin, the vehemence of anthropologist anti-Africa sentiment became to decline, given the vast numbers of early human fossils that have been recovered in the continent in recent years. The episode is also a reminder that scientists are too often guided as much by emotion as by reason. (It was emotion wrapped with racism that prevented early investigators from exploring the African continent, let me add.)

In Darwin's second major conclusion in *The Descent of Man*, he argued that the evolution of our unusual mode of locomotion was directly linked to the manufacture of stone weapons as used by apes. He went further and linked these evolutionary changes to the origin of the canine teeth in humans, which are unusually small compared to the dagger-like canines of apes. "The early forebears of man (apes) were probably furnished with great canine teeth," he wrote in the Descent of Man; but as they gradually acquired the habit of using stones, clubs, and other weapons for fighting with their enemies or their rivals, instead of their teeth, they would use their jaws and teeth less and less for fighting. In this case, the jaws, together with the teeth, would become reduced in size," Darwin concluded.

These weapon-wielding, bipedal (standing on two legs) creatures developed a more intense social interaction which demanded more intellect, Darwin argued further. And the more intelligent our ancestors became, the greater was their technological and social sophistication, which in turn demanded an ever-larger intellect. And so on, as the evolution of each feature fed on others. This hypothesis of linked evolution was a very clear scenario of human origins, and it became central to the development of the science of anthropology by Darwin's believers and supporters.

According to this scenario, the original human species was more than merely a bipedal ape; it already possessed some features we value in Homo Sapiens. The image was so powerful and plausible that anthropologists were able to weave a persuasive hypothesis around it for a very long time. But the scenario went beyond science. If the evolutionary differentiation of humans from apes was both abrupt and ancient, a considerable distance was inserted between us and the rest of nature. For those with a conviction that Homo Sapiens is a fundamentally different kind of creature, this view point offered comfort.

Such a conviction was common among scientists in Darwin's time, and well into this century, too. For instance, the 19th century English naturalist Alfred Russell Wallace, who also invented the theory of natural selection independently of Darwin, balked at applying the theory of those aspects of humanity we most value to apes. He considered humans too intelligent, too refined, too sophisticated to have been the product of mere natural selection, evolving from the ape world to our modern form. Russell reasoned that primitive hunter-gatherers would have no biological need for these qualities, and so could not have arisen by Darwin's definition and application of natural selection. Supernatural intervention, he felt, must have occurred to make humans so special. Wallace was not alone. So were many scientist and thinkers before him. However, Wallace's lack of conviction in the power of natural selection greatly upset Darwin. Current evidence shows that Wallace was right.

The Scottish paleontologist, Robert Broom, whose pioneering work in South Africa in the 1930s and 1940s helped establish Africa as the cradle of mankind, also expressed strong views on human distinctiveness. He believed that Homo Sapiens was ultimate product of evolution and that the rest of nature had been shaped for his comfort. Like Wallace, Broom looked to supernatural forces in the origin of our species. These two scientists and many more during Darwin's time accepted that fact that Homo Sapiens derived ultimately from nature through the process of evolution, but their belief in the essential spirituality, or transcendent essence, of humanity led them to construct sensible explanations for evolution which maintained human distinctiveness. They rejected Darwin's "evolutionary package."

Darwin's argument remained influential until a little more than a decade ago, and was effectively responsible for a major dispute over when humans first appeared. Further investigations on this question also marked the end of its sway over anthropological thinking. Man as Darwin thought did not evolve from African apes

Dr. Nyonbeor A. Boley Sr.

Africa: The Continent of Our Human Origin

Rick Potts

Writing about the evolution of the human race based on real science, Dr. Rick Potts wrote: 'Human revolution is the lengthy process of change by which people originated from apelike ancestors, not apes. Scientific evidence shows that the physical and behavioral traits shared by all people originated from apelike ancestors and evolved over a period of approximately six million years.'

One of the earliest defining human traits, bipedalism—the ability to walk on two legs—evolved over 4 million years ago. Other important human characteristics—such as a large and complex brain, the ability to make and use tools, and the capacity for language—developed more recently. Many advanced traits—including complex symbolic expression, art, and elaborate cultural diversity—emerged mainly during the past 100,000 years.

Humans are by definition primates. Physical and genetic similarities show that modern human species, Homo sapiens, has a very close relationship to another group of primates species, the apes. Humans and the great apes (large apes) of Africa—chimpanzees (including bonobos, or so-called "pygmy chimpanzees") and gorillas—share a common ancestor (genetically) that lived between 8 and 6 million years ago. Humans first evolved in Africa, and much of human evolution occurred on that continent. The fossils of early humans who lived between 6 and 2 million years ago come entirely from the continent of Africa.

Most scientists currently recognize some fifteen to twenty different species of early humans. Scientists do not all agree, however, about how these species are related or which ones simply died out. Also, scientists debate over how to identify and classify particular species of early humans, and about what factors influenced the evolution and extinction of some species. But the fact must be stressed here—they all evolved out of Africa.

Early humans first migrated out of Africa into Asia probably between 2 million and 1.8 million years ago. They entered Europe somewhat later, between 1.5 million and 1 million years. Species of modern humans populated many parts of the world much later. For instance, people first came to Australia probably within the past 60,000 years and to the Americas within the past 30,000 years or so. The beginnings of agriculture and the rise of the first civilization as defined by the West occurred within the past 12,000 years.

The process of evolution involves a series of natural changes that cause species (populations of different organisms) to arise, adapt to the environment, and become extinct. All species or organisms have originated through the process of biological evolution. In animals that reproduce sexually, including humans, the term species refers to a group whose adult members regularly interbreed, resulting in fertile offspring—that is offspring themselves capable of reproducing. Scientists classify each species with a unique, two-part scientific name. In this system, modern humans are classified as Homo Sapiens.

Evolution occurs when there is a change in the genetic material—the chemical molecule, DNA—which is inherited from the parents, and especially in the proportions of different genes in a population. Genes represent the segments of DNA that provide the chemical code for producing proteins. Information contained in the DNA can change by a process known as mutation. The way particular genes are expressed, that is, how they influence the body or behavior of an organism, can also change. Genes affect how the body and the behavior of an organism develop during its life, and this is why genetically inherited characteristics can influence the likelihood of an organism's survival and reproduction.

Evolution does not change any single individual. It does not change an ape to man, for example. Instead, it changes the inherited means of growth and development that typify a population (a group of individuals of the same species living in a particular habitat). Parents pass adaptive genetic changes to their offspring, and ultimately these changes become common throughout a population. As a result, the offspring inherit those genetic characteristics that enhance their chances of survival and ability to give birth, which may work well until the environment changes. Over time, genetic change can alter a species' overall way of life, such as what it eats, how it grows, and where it can live. Human evolution took place as new genetic variations in early ancestor population favored new abilities to adapt to environmental change and so alter the human way of life." Humans come from humans. Nothing more and nothing less.

Darwin's evolutionary theory was a mechanistic theory of life which would favor the well-adapted, the fittest, and would eliminate all other organisms. As far as Charles Darwin was concerned man has no special status other than his definition as a distinct species of animals. He is in the fullest sense a part of nature and not apart from it. Man is akin, not figuratively but literally, to every living thing, be it amoeba, a tapeworm, a flea, a sea worm, an oak tree, or a monkey, even though degrees of relationship are different.[19] This is what Darwin would have you believe. Is it not strange that so many biologists believe in this kind of nonsense, even though Charles Darwin himself had a change of mind in his final days.

> "Darwin's evolutionary theory did not tell the full story."
>
> - His Holiness, The Dalai Mala.

One of the world's most powerful thinkers of our time, His Holiness, The Dalai Lama does not accept Darwin's evolutionary theory as far as the evolution of human beings or sentience beings is concerned. His Holiness states, "on the whole, I think the Darwinian theory of evolution, at least with the additional insights of modern genetics, gives us a fairly coherent account of the evolution of human life on earth. However, despite the success of the Darwinian narrative, I do not believe that all the elements of the story are in place. To begin with, although Darwin's theory gives a coherent account of the development of life on this planet and the various principles underlying it, such as natural selection, I am not persuaded that it answers the fundamental question of the origin of life. Darwin himself, I gather, did not see this as an issue. Furthermore, there appears to be certain circularity in the notion of "survival of the fittest." The theory of natural selection maintains that, of the random mutations that occur in the genes of a given species, those that promote the greatest chance of survival are most likely to succeed. However, the only way this hypothesis can be verified is to observe the characteristics of those mutations that have survived. So in a sense, we are stating simply this: 'Because these genetic mutations have survived, they are the ones that had the greatest chance of survival."

From the Buddhist perspective, the idea of these mutations being purely random events is deeply unsatisfying for a theory that purports to explain the origin of life. Karl Popper once commented that, to his mind, Darwin's theory of evolution does not and cannot explain the origin of life on earth. For him, the theory of evolution is not a testable scientific theory but rather a metaphysical theory that is highly beneficial for guiding further scientific research. Moreover, the Darwinian theory, while acknowledging the critical distinction between inanimate matter and living organisms, fails to acknowledge adequate qualitative distinctions between living organisms, such as trees and plants on the one hand and sentient creatures on the other hand.

One empirical problem in Darwin's focus on the competitive survival of individuals, which is defined in terms of an organism's struggle for individual reproductive success, has consistently been how to explain altruism, whether in the sense of collaborative behavior such as food sharing or conflict resolution among animals like chimpanzees or act of sacrifice. There are many examples, not only among human beings but among other species as well, of all individuals who put themselves in danger to save others. For instance, a honeybee will sting to protect its hive from intruders, even though the act of stinging causes it to die; the Arabian babbler, a type of bird, will risk its own safety to warn the rest of the flock of an attack.

Post-Darwinian theory has attempted to explain such phenomena by arguing that there are circumstances in which altruistic behavior, including self-sacrifice, enhances an individual's chances of passing on its genes to future generations. However, I do not think this argument applies to instances where, I am told, altruism can be observed across species. For example, one might think of the host birds who wean and rear baby cuckoos left in their nests, although some have explained this exclusively in terms of the selfish benefit obtained by the cuckoos. Moreover, given that this kind of altruism does not always appear to be voluntary—some organisms seem programmed to act in a self-sacrificial way—modern biology would basically see altruism as instinctive and dictated by genes. The problem becomes all the more complex if we bring in the question of human emotion, especially the numerous instances of altruism in human society.

Some more dogmatic Darwinians have suggested that natural selection and survival of the fittest are best understood at the level of individual genes. Here we see the reduction of the strong metaphysical belief in the principle of self-interest to imply that somehow individual genes behave in a selfish way. I do not know of how many of today's scientists hold such radical views. As it stands, the current biological model does not allow for the possibility of real altruism.

Again, His Holiness states, "At one of the Mind and Life conferences in Dharamsala, India, the Harvard historian of science Anne Harrington made a memorable presentation on how, and to some extent why, scientific investigation of human behavior has so far failed to develop any systematic understanding of the powerful emotion of compassion. At least in modern psychology, compared with the tremendous amount of attention paid to negative emotions, such as aggression, anger, and fear, relatively little examination has been made of more positive emotions, such as compassion and altruism." This emphasis may have risen because the principle motive in modern psychology has been to understand human pathologies for therapeutic purposes. However, I do feel that it is unacceptable to reject altruism on the grounds that selfless acts do not fit within the current biological understanding of life or are simply redefinable as expressions of the self-interest of the species. This stance is contrary to the very spirit of scientific inquiry. As I understand it, the scientific approach is not to modify the empirical facts to fit one's theory; rather the theory must be adapted to fit the results of empirical inquiry. Otherwise it would be like trying to reshape one's feet to fit the shoes.

I feel that this inability or unwillingness to fully engage the question of altruism is perhaps the most important drawback of Darwinian evolutionary theory, at least in its popular version. In the natural world, which is purported to be the source of the theory of evolution, just as we observe competition between and within species for survival, we observe profound levels of cooperation (not necessarily in the conscious sense of the term). Likewise, just as we observe acts of aggression in animals and in humans, we observe acts of altruism and compassion. Why does modern biology accept only competition to be the fundamental operating principle and only aggression to be the fundamental trait of living beings? Why does it reject cooperation as an operating principle, and why does it not see altruism and compassion as possible traits for the development of living beings as well?

To what extent we should ground the entirety of our conception of human nature and existence in science depends, I suppose, on what conception of science we hold. For me, this is not a scientific question; rather it is a matter of philosophical persuasion. A radical materialist might wish to support the thesis that evolutionary theory accounts for all aspects of human life, including morality and religious experience, while others might perceive science as occupying a more limited scope in understanding the nature of human existence. Science may never be able to tell us the full story of human existence or even, for that matter, to answer the question of the origination of life on earth. This is not to deny that science does, and will continue to, have a lot to say about the origination of the tremendous diversity of life forms. However, I do believe that as a society we must accept a degree of humility toward the limits of our scientific knowledge of ourselves and the world we live in.

If twentieth-century history, with its widespread belief in social Darwinism and the many terrible effects of trying to apply Eugenics that resulted from it, has anything to teach us, it is that we humans have a dangerous tendency to turn the visions we construct of ourselves into self-fulfilling prophecies. The idea of the "survival of the fittest" has been misused to condone, and in some cases to justify, excesses of human greed and individualism and to ignore ethical models for relating to our fellow human beings in a more compassionate spirit. Thus, irrespective of our conceptions of science, given that science today occupies such an important seat of authority in human society, it is extremely important for those in the profession to be aware of their power and to appreciate their responsibility. Science must act as its own corrective to popular misconceptions and misappropriations of ideas that could have disastrous implications for the world and humanity at large. (Dalai Lama, 2013)

Regardless of how persuasive the Darwinian account of the origins of life may be, as a research scientist, I find it leaves one crucial area unexamined. This is the origin of sentience —the evolution of conscious beings who have the capacity to experience pain and pleasure. After all, from Buddhists' and natural historians' perspective, the human quest for knowledge and understanding of one's existence stems from a profound aspiration to seek happiness and overcome suffering. Until there is a credible understanding of the nature and origin of consciousness, the scientific story of the origins of life and the cosmos will not be complete.

Thus, His Holiness, the Dalai Lama could not accept Darwin's claims of the origins of life on earth. When untestable claims are accepted as fact, the spirit of scientific inquiry is in jeopardy. And as you will see in the next chapter, Charles Darwin had trouble believing it himself.

The Church of England and the Vatican Silence on "Evolution"
As Darwin himself later indicated, "There is no necessary conflict between evolutionary biology and Biblical creationism. Only the insistence on strict biblical literalism, the belief that everything in the Bible must be true, including the mutually inconsistent ac-

counts in Genesis on the origin of the Earth and life within six days and the scientific impossibility of putting each pair of species into a boat, forces a collision between the word of science and the Judeo-Christian religion." Could this be the real reason why Charles Darwin wrote his "evolutionary theory"? To counter biblical explanations? Was this theory an attempt to justify his rebellion against an "all powerful Diety"? Essentially, Darwin did not support his own "theory" in his final day as a scientific or biological possibility. Charles Darwin's evolutionary "theory" from the point of view of investigative science is not a theory but a proposition or hypothesis. Current research in science, particularly, molecular biology justifies the conclusion reached in this tiny book that Darwin's "theory" is not a scientific theory. But yet, the powerful Church of England and the Vatican had nothing to say about Darwin's publication on the origin of man! Another interesting question remains, given the fact that Charles Darwin did not have the science to back his "theory." Why has his theory influenced so many well-meaning scientists? Were these scientists fed up with Church domination over issues of science, religion, and God and decided to follow Darwin's views on evolution as a challenge to the Church?

It should be recalled, for example, of a story told of a philosopher-scientist, Nicholas Copernicus who dared to propose the heliocentric hypothesis that the Sun is the center of the universe as opposed to Church's doctrine, favoring Earth as the center of the universe. At first, because of difficulty with mathematical calculations being extremely complicated, the revived heliocentric hypothesis did not make any significant dent in the prevailing cosmology. Besides, you just had to look up into the sky and you could see that the Sun was moving, while the Earth was "obviously" motionless. Because of the intractable complexity of the planetary motions as viewed from Earth, however, Copernicus' view gradually gained influential supporters who spread his ideas, and nearly seventy-five years after their first publication, Pope Paul V decided it was time for the Church to take another look at them. So in 1616, he convened a theological panel that unanimously agreed that heliocentricism was "formerly heretical." Copernicus' book was withdrawn, pending corrections to it by the Church. The Church had halted scientific inquiry at this time, however brief.

One of Copernicus's most famous champions at this time was Galileo, and he immediately worried that he himself might be charged with heresy. Sixteen years earlier, his well learned countryman, the Pantheistic mystic and former Dominican friar, Giordano Bruno, had been burned at the stake by the Church for his Copernican support and other heretical views. Galileo had already caused a stir when he published the "Starry Messenger" in 1610; it was an account of new observations he had made using his remote telescope, a recent Dutch invention Galileo had personally improved upon. In particular, his discovery that Jupiter has moons, which orbit it rather than Earth, upset the Earth-centered view of the universe held by the Church. But the book, whose first edition sold out in a few days, was passed by censors of the Roman Inquisition. After all, it contained no evidence that the Earth itself moved. Now, however, Italian scientists were unnerved by the anti-Copernicus edict of 1616.

So in 1623, a new pope, Urban VIII, took office; he was unusually interested in the emerging scientific view of the world and had long been an admirer of Galileo. He personally encouraged him (Galileo) to keep on with his work as long as he did not defend Copernicus' hypothesis as a proven fact. But in 1632, Pope Urban felt personally betrayed when Galileo published his pro-Copernican book *Dialogue Concerning Two Chief World Systems*, (that is, the Earth-centered system favored by the Church, and which had been handed down from the ancient Greek astronomer Ptolemy and Copernicus' Sun-centered system). Brought before the Inquisition, Galileo argued that *Dialogue* was simply a debate—a dialogue—but it was evident that he favored the heliocentric side of the argument and he was charged with heresy in 1634.

Approaching seventy, and after a lifetime of service to science, he was absolved of the usual "censures and penalties" for the crime by agreeing formally, and humiliatingly, to deny that the Sun was the centre of the world, to recite the seven penitential psalms' once a week for the next three years, and to refrain from all teaching and other public scientific activities. He was helped in this last matter by being sentenced to imprisonment in his own home at the Inquisition's pleasure. Meanwhile, the *Dialogue* was banned; it remained officially prohibited by the Catholic Church until 1822. Charles Darwin was eleven years old in 1822 when this prohibition ended. Galileo was formally absolved in 1922. Charles Darwin had been *dead 40 years by then, meaning that Darwin lived throughout the period in which Galileo's work was banned. Given the authority of the Church of England and the Vatican before and during Darwin's time, it is* tempting to believe that scientists decided to uphold Darwin's unsubstantiated claims such as that of the evolutionary hypothesis simply to anger the Church as revenge. As indicated above, priests have killed many scientists! According to Church doctrine, God created man in his own image. Evolutionary hypothesis by Darwin indicated otherwise, that in fact, God did not create man and that man evolved from lower animals, including tapeworms and apes. Because of Darwin's social status, I suppose, the Church of England nor the Vatican did not question his publication nor punished him. Historically, it is a well established fact that priests have killed many philosophers and scientists who taught doctrines opposed to Church's teachings as demonstrated in the passage above. Why was Darwin's "theory on the origin of life" treated differently? Was this Charles Darwin's view all along? Did he always believe that God did not create the world and the human beings in it? Here, we review Darwin's views about religion, particularly God and the origin of man. This, you will see in the next chapter.

Chapter Four
Charles Darwin and Religion: As He Recant

Indeed, Charles Darwin's theory of "evolution" forcefully created intense debates about God, religion, and science ever since it was published, giving more ammunition to those who were already tired and frustrated with the Church of England and the Vatican domination over issues of God, religion, and science. So it is worthy to consider what his own religious beliefs were before the publication of *Descent of Man* which is the centerpiece of his paper published in 1871. What was in his mind at the time of his writing? What caused Charles Darwin to change his mind about God and the origin of man? Is the publication of the evolutionary theory Darwin's revolt against his previously held belief about God and the origin of man and the other species? Was Charles Darwin for creationism before he was against it? It is imperative to consider what his own religious views were because, just as "his theory has influenced people's views about God, his view of God has helped to shape his theory."[7]

Many people in our time will be surprised to learn that Charles Darwin, as a young man, attended church with his mother and received religious training at a Catholic Church of England boarding school. Darwin even attended Christ's College, now Cambridge University, and studied theology, earning a B.A. degree in Theology (1827-1831) to prepare for life as a country parson, the equivalent of priesthood today. Charles Darwin even confessed at some point in time, especially shortly after his graduation, saying that "he did not doubt the strict and literal truth of every word in the Bible." In fact, he wrote in his autobiography that he was at one point led by the conviction of the existence of God, the creation of the world in six days, and of the immortality of the soul," believing that "there is more in man than the mere breath of his body." At this point, it is fair to say that Charles Darwin was a practicing Christian and therefore a Creationist. If the soul of man is immortal then man did not evolve from a molecule!!

Charles Darwin recalled that at the initial time of writing *On the Origin of Species* he was convinced of the existence of God as an intelligent "First Cause, Creator of all that exists," and deserved to be called a theist. However, his view drastically changed while on board the Beagle and by the time he returned to England in 1836 he had come to view God as a "revengeful tyrant." We now wonder, what happened to Darwin? There is indication that during the five-year Beagle voyage he had ample opportunity to see the cruelties of slavery and wondered how and why a powerful God would allow such inhumanity to exist. He wondered why an all-knowing, all powerful God would allow such magnitude of suffering to occur. For this reason he could not accept that a kind God would allow men to live in such a wretched state as many of the native people he saw during this voyage. The issue of why God would allow slavery and suffering in the world was an internal conflict that Darwin could not resolve. He recorded the thoughts he harbored at that time against God. He wrote:

A being so powerful and so full of knowledge as the God who could create the universe in six days, is to our finite minds omnipotent and omniscient, and it revolts our understanding to suppose that His benevolence is not unbounded....This whole argument from the existence of an intelligent First Cause, all powerful deity seems to me a strong one[2].

Crowning this confusion and the issue of suffering was the death of Charles Darwin's beloved ten year old daughter, Annie Elizabeth Darwin in 1851. This tragedy dealt a crushing blow to his religious beliefs, as Darwin began to deliberate about the Christian meaning of mortality and lost all faith in a beneficent, all knowing, all powerful God who could not save his daughter's life. Of course, at this time in his life, however, he continued to give support to his local parish or church, but on Sundays would go for a walk while his family attended Masses. Darwin therefore reasoned that death and suffering were integral to the operation of the world and had always existed. His doubt in a personal God who is beneficent, answers prayers, parts the waters, and is involved in the affairs of man totally evaporated! The stage is set for a revolt or rebellion against an all powerful and all beneficent God who couldn't save the life of his beloved daughter. The death of Darwin's ten year old daughter, Annie, in 1851 is widely believed to have removed whatever lingering religious sentiment he may have had by then. Annie's death also helped steel his resolve at long last to publish his theory of "evolution," that in fact, God did not create the world. Nevertheless, he still professed a residual belief in God, the Almighty.

In a letter to his friend, an American botanist, Asa Gray in 1860, nine years after the death of his daughter, Charles Darwin still acknowledged that God was the ultimate lawgiver and intelligent designer, but he could not see an omnipotent deity in all the pain and suffering in the world, not certainly a God who is concerned about the personal affairs of the human race. He wrote:

> I have no intention to write in the manner and fashion of an atheist, but I own that I cannot see as plainly as others do, and as I should wish to do, evidence of beneficence on all sides of us. There seems to me too much misery in the world… On the other hand, I cannot anyhow be contented to view this universe, its wonderful organization, and especially the nature of man, and to conclude that everything is a result of Brute Force, or Big Bang. I am inclined to look at everything as resulting from designed natural laws, with the details, whether good or bad, left to the working out of what we may call chance.

Still searching for support for his theory of evolution, Darwin contacted another well known scientist of his day. Swiss-born paleontologist, Louis Agassi of Harvard University was one of the best trained scientists of his age, and he knew the fossil record better than any man alive. Hoping to enlist Agassi as an ally since Gray did not accept his theory, Darwin sent him a copy of the *Origin Of Species* and asked him to consider the argument with an open mind. One can almost imagine the great naturalist receiving the unremarkable package from the postman, quickly unwrapping the small green volume that had stirred such a tempest on both sides of the Atlantic—Europe and the United States. He read the book with deep interest, making notes in the margins as he moved through it, but in the end his verdict would disappoint its author, Mr. Darwin. Agassi concluded that the fossil record, particularly the record of the explosion of Cambrian animal life, posed an insuperable difficulty for Darwin's theory (page 8).

In summary, the Cambrian fossil evidence represented a significant challenge to Darwin's claim that Natural Selection had the capacity to produce novel forms of animal life. Agassi thought the evidence of abrupt appearance, the absence of ancestral forms in the Cambrian Explosion, refuted Darwin's theory completely. "Of these earlier forms, where are their fossilized remains?" the great scientist wondered.

For the benefit of those who have not read about the Cambrian Explosion, a brief explanation is given below.

Between 570 and 530 million years ago, all the basic body plans found in nature today abruptly appeared during this time: bodies with heads, animals with tails, and appendages, all specialized segments performing specialized functions. All animal evolution for the last half billion years has come from tinkering with these Cambrian body plans, however primitive in their forms. Simply put, a complex biological structure with many interacting parts just appeared, at first glance, as if it were originally created in its present form with all its interlocking components fully formed and intact. It doesn't seem possible that they developed step by step via biological evolution. Isn't this the death of Darwin's theory of the origins of species by natural selection? As you continue to read you will appreciate the Legends of the Konobo people on the origin of man and all animals and plants. Their oral stories are closely related to the Cambrian Explosive Era.

As he developed his theory on the origins by purely natural means, he grew further from the biblical concept of a Creator and said of his religious views, as if speaking to an individual, "I am sorry to have to inform you that I do not believe in the Bible as a divine revelation, and therefore, not in Jesus Christ as the Son of God." [4] He came to think that the religious instinct had evolved with society and eventually concluded, "For myself, I do not believe that there had been any revelation. As for a future life, every man must judge for himself between conflicting vague probabilities." Charles Darwin, in his final years on Earth became agnostic entirely and was no more a Christian. In this context, Darwin was not alone. So were Spinoza, Leo Tolstoy, Albert Einstein, Bernard Haisch, Stephen Hawkins, Ghandi, and many others as we shall see later.

Departing from his Christian faith, Darwin spent the rest of the decade thinking evolution through and finally coming up by 1838 with the central mechanism of his theory: Natural Selection. [Eldredge, p-74] The question may then be asked: Why had

Charles Darwin waited twenty years, despite his ambition to "take his place among men of science?" Darwin was by academic training a theologian, not a scientist..But why did he wait so long to publish his theory? Apparently, Darwin knew his idea was truly earth-shattering and would, as he foresaw it, transform the way science looked at life at that time and would anger members of the elite or imperial class who were all creationists. If God did not create the universe and the human beings in it, who did? To avoid opposition from the ruling class and the Church, Darwin had to insert or include "Natural Selection" in his publication. Darwin clearly remembered what had happened to Socrates, Copernicus, Bruno, and Galileo who went against established views. The field of paleontology had not advanced by then. Neither had molecular biology. Darwin did not study fossil remains extensively to have come up with evidence for the origin of species, including man. His hypothesis of the origin of species was inaccurate.

It was that stabilist, elitist point of view, the very one he grew up in and benefited from, that caused Darwin to delay the publication of his "theory." Charles Darwin knew that to do the job right, to avoid the wrath of his family members and the ruling class, he had to have ironclad evidence and a convincing explanation of how evolution happens and the role of God in the creation of humankind relative to the imperial establishment. The real key to Darwin's dilemma lies in the fact that most of British Society adhered to the biblical view of the origin of the world and of life—and the stability of all things more or less since their creation, as outlined and upheld especially by the Church of England

and the Vatican in Rome. This encompassed people from all walks of life, including, critically, the academic world. Everyone or nearly everyone was a creationist. In fact, there was little in the way of a professional class of scientist when Darwin came along, though by midcentury such an enlightened class had begun to emerge. No one dared challenge the doctrines of the Church, whether in science or medicine.

Darwin's mentors and older colleagues from the 1820s into the 1850s were either the very rich elite or clergymen. All religious, all vested in the status quo. And the status quo meant first and foremost stability, no uprising, no revolutionary or strange ways of looking at things. The upper class was favored by a vestigial form of "divine right." They deserved to be the elite because that was the way things were, the way God wanted it. [Eldredge, p-76] The idea of "Natural Selection" was the new way of justifying the existence of the elite: the thought that cream rises to the top and that the people with all the money and power occupying the highest social strata are there because they deserve to be because God created it that way.

So Darwin was not just against a prevailing scientific view, for the scientific view, echoing the commonsensical perception that the world and its species is indeed stable, fit in so well and nicely with the prevailing religious views to the extent that to attack one is to attack the other. Indeed, the "scientific view" was essentially a religious view at that time. The rising tide of rationalism or reasoning in science, such achievements as Newtonian physics and the beginning of chemistry, had yet to influence ideas on the living world. Religious doctrine dominated biological thinking. [Eldredge, p-78] That was the case when Darwin published his "theory." The Church would punish anyone who taught against established religious doctrines on creation and "Natural Selection." Imperialism, the supremacy of the clergy, was according to God's divine will as Darwin saw it.

That is why Darwin's face is on the ten-pound note in Great Britain today. He had to be rewarded. Even though creationism is a less vociferous strain in modern British life than it is in many of her former colonies (United States of America, included), in his native England there are many people who are still uncomfortable with the anti-religious nature of the idea of evolution and natural selection.[9] The queen and king of present-day Great Britain still live the life of natural selection and the preservation of a favored people. They are the elite, imperial power, and naturally selected!! In summary, Darwin's "theory" does not describe the origin of life on Earth and our place in it based on science. Darwin's so-called theory of evolution is therefore voodoo science. The advance of molecular biology and our understanding of science make it impossible to consider Darwin's views seriously.

In the world of Charles Darwin man has no special status other than his definition as a distinct species of animals. He is a bag of chemical molecules running around planet Earth. But this is factually untrue. We know man has a special status in this world. There are still apes, monkeys, tapeworms, and amoeba still living in their primitive environments as I write these words.

Dr. Paul Weis, one of the founders of biological system theory, writing in his essay *The Living System: Determinism Stratified* presented his attitude towards the living world and those who think that human beings evolved from molecules: "Biology must retain the courage of its own insights into the living nature; for, after all, organisms are not just heaps of molecules. At least, I cannot bring myself to feel like one. Can you? If not, my essay may, at any rate, have given you some food for thought." (Weiss, 1969, p-400)

I have been very, very attracted by this conclusion, and I admit that I actually cannot bring myself to feel and act like a heap of molecules, although I am aware that my body is indeed composed of molecules and that even my thinking and feeling is traced back to molecular structures and functions. But I am more than my physical structure, or a molecular composition. I am a spiritual being temporarily housed in this molecular structure.

Part II

"What is the meaning of human life, and for that matter, of the life of any creature? To know an answer to this question is to be religious. The man who regards his own life and that of his fellow creatures as being meaningless is not only unhappy, but hardly fit for life."

Albert Einstein

Chapter 5
Natural Theology: To the Rescue

Charles Darwin's main dilemma was the inevitable consequence of pursuing a system of humanistic and rational enquiry like science that seeks to explain all apparent mysteries in terms of lawful, natural phenomena at the time when his own past experience and belief indicated something else. He was forced to choose an intellectual path between a belief that the world was created exactly as it is now in six days and probably some 6,000 years ago and a growing evidence that the universe could or might have been created but that such creation is mysterious. In fact, in various guises and under different names, what Darwin described as evolution had been in the air for around two hundreds before the publication in 1859 of his theory on the *Origin of Species by Means of Natural Selection and the Preservation of Favored Races in the Struggle for Life* as clearly stated in the introduction. Since 1650, the development of a whole range of new sciences had posed threat after threat to the mysteries of revealed religion by so-called "inspired men." Evolution is part of a broader and older enquiry and a deeper contest for our intellectual commitment, a contest between a world system that expects every part of the cosmos ultimately to be explainable in terms of natural properties and one that maintains the existence of a fundamental core of unknowability, of supernatural mystery, and the controlling hand of an eternal non-worldly Being. This, indeed, is mankind's oldest intellectual puzzle that requires reason, not revelation, to unlock. Darwin's attempt to participate in this intellectual and scientific discourse is appreciated but his conclusion about the origin of humans and other species is simply wrong.

Darwin's "theory" of the origin of species was generally seen as an attack on the Christian religion. This was not because it contradicted the story of creation in the Bible. By the mid-1800s, many educated people in Britain, Europe, and the United States had begun to view the Genesis story in a new way. They saw it not as an actual account of creation but as a story symbolizing God's power. All the same, faithful Christians and even people with liberal religious beliefs were profoundly shocked that God seemed to have no role in the world Darwin envisioned[2].

At the same time, many religious thinkers accepted the idea of natural theology (defined below). The chief proponent of natural theology was a British clergyman named William Paley, who had written an influential book published in 1802, long before Darwin's publication on the *Origins of Species*. Paley saw God not as a hands-on Creator of the Genesis story. Instead, God was a divine authority who set the world in motion and established natural laws that keep it running.

As proof of God's existence and role in creation, Paley pointed to the evidence of design in the world. He used the image of a watch to demonstrate his point. He wrote that when we examine the complex mechanism of a watch, we understand that it has been made for a specific purpose. From that, we assume that some individual or being designed the watch. Similarly, Paley reasoned that all the complex elements of the natural world represent the work of a divine designer, God or Spirit [2]. Darwin argued otherwise.

In Darwin's "theory," the intricate forms of natural organisms illustrate not the work of God but of natural selection. It is this force that creates complex beings by selecting and combining random variations. Darwin's "theory" does not rely on a vision of grand design with a divine designer.

The idea of a world with no divine plan or purpose was unthinkable to many, including some of Darwin's friends and supporters. One such supporter was Asa Gray, a botanist from Harvard University. When pressed for further evidence by Asa Gray, Darwin had to summersault. Darwin wrote to Gray saying, "I cannot think that the world, as we see it, is the result of chance; & yet I cannot look at each separate thing as the result of design." Finally, in the second edition of his book, Darwin had to make some changes by adding a notion of a creator to the final paragraph: "There is a grandeur to this view of life, with its several powers, having been originally breathed by the Creator into a few forms of many…"[2]

It has not always been necessary to choose between one side or the other—science or religion, reason or revelation. In the age before Charles Darwin, many powerful clerics were notable scientific scholars and leading scientists in their own right. For them science and religion could share a common philosophical basis with the premise that a careful, rational study of nature, instead of denying the existence of God, would confirm that all life is, after all, the product of God's unique creation. This was the view of scientists including Ernst Mach, Alfred North Whitehead, Herbert Wildon Carr, Albert Einstein, among many others—the true approach of natural theology.

Natural theology is by definition a branch of theology based on reason and ordinary experience, or a program of inquiry into the existence and attributes of God without referring or appealing to any divine revelation. Natural theology rejects knowledge based on revelation.

Natural theology and its counterpart in the geological context, physico-theology, provided an intellectual framework that both embraced science and kept it at bay. Indeed, natural theologians believed that a study of God's handiwork constituted a proof of the very existence of God. Believers who were scientists and philosophers welcomed natural theology because it gave their endeavors a framework within which to operate. Deist and Christian alike could find much to favor in a movement that sought to discover God through rational study without depending on a belief in miracles and revelation, or insisting on the literal truth of the Bible and the many inconsistent and impossible events it contains. For instance, many natural theologians could not believe the story of Adam and Eve, the "Flood" with the many different species known to science being harbored in one boat, Noah and his cousins living well over 500 years of age, including Methuselah who reportedly lived 969 years old! Because of this distinction or approach by natural theologians, most mainstream Christian scholars began to worry that such a new movement would risk flirting too seductively with material and science-based explanation of the world and preferred to remain with the relative safety of the authority of the Bible and revelation. In fact, St. Augustine, Bishop of Hippo, an esteemed authority of his day proclaimed that God had already prepared a place called "Hell" to punish those who question biblical authority, including creation and the associated events. Thus, mainstream Christian scholars went their separate way with the belief that every word in the Bible is literally the word of God and must be obeyed as such.

Nevertheless, in one form or another, natural theology has maintained a currency to our present day, and its concept is spreading slowly but surely. Natural theology affords a starting point from which to trace a story that reaches from the ancient Greeks to Descartes and to the intellectual environment of the nineteenth and twentieth century from which Edmund Ware Sinnott, Albert Einstein, Stephen Hawking, and Bernard Haisch sprang. Edmund Ware Sinnott, a prominent biologist at Yale University, writing in his book *Two Roads to Truth* attempted a reconciliation between science and religion based on the result of contemporary science or natural theology. He writes:

After the revolution introduced by relativity, quantum mechanics, and nuclear physics, science was forced to modify some of its earlier conclusions. The plain truth is that the universe is a much more complex system than it seemed to be in Newton's time…Scientists accept now without surprise ideas that would have seemed preposterous not long ago. This change has been reflected in a more open-minded attitude on their part towards idealistic philosophies. For three centuries a confidently advancing science seemed to undermine the very foundation of faith, and religion was forced to modify its position in many ways or lose the support of its more thoughtful partisans. The tide, however, has begun to turn, and an aggressive idealism is going from the defense to the attack.

Those are the natural theologians. For their part, mainstream Christians have refused to modify their story of creation and of the origin of man despite the scientific or commonsense impossibility of these events.

If mainstream Christianity must survive as it should, it must make many adjustments and makeovers. The time has come for Christians to revisit many claims of the past.

Albert Einstein, the preeminent theoretical physicist of his day, shared the conclusions of natural theology as many other real scientists do. The belief in the mysterious, the conviction that man did not evolve from lower animals or from a lifeless molecule by chance caused Albert Einstein departure from mainstream theoretical physics of his time when it became fashionable to consider man as nothing other than a bag of molecules and the creation of the universe as a matter of chance.

Quantum mechanics introduced an unavoidable element of unpredictability or randomness into science that Einstein could not accept. Einstein objected to this view of Quantum mechanics very strongly despite the remarkable and pioneering role he played in the development of quantum theory. He never accepted the idea that the universe was governed by chance; his feelings were summed up in his famous statement, "God does not play dice." (p. 73 – A Brief History of Time, the Universe in a Nutshell- by Stephen Hawking).

Nevertheless, Einstein was awarded the Nobel Prize for his contribution to quantum theory—the photo-electric effect.

Although skeptical about traditional philosophy and a belief in a personal God, Albert Einstein had the deepest respect for the mysteries posed by true religion, especially the nature of existence, the universe, and our place in it. He once wrote: "Science without religion is lame, religion without science is blind." He would also attribute this appreciation of mystery as the source of all science: "All the fine speculations in the realm of science spring from a deep religious feeling." Einstein wrote: "The most beautiful and deepest experience a man can have is the sense of the mysterious. It is the underlying principle of religion as well as all serious endeavors in art and science." He concluded, "If something is in me which can be called religious, then it is the unbounded admiration for the structure of the universe so far as science reveals it." Perhaps his most elegant and explicit statement about religion, God, and creation was written in 1929. When asked if he was a Christian or if he believed in God, he said, "I am not an atheist and I don't think I can call myself a pantheist." Regarding creation he said, "We are in the position of a little child entering or finding himself in a huge library filled with books written in many different languages. The child knows someone must have written those books. It doesn't know how. It does not understand the languages in which they are written. The child dimly suspects a mysterious order in the arrangement of the books but it doesn't know what it is. That, it seems to me, is the attitude of even the most intelligent human being toward God. We see a universe marvelously arranged and obeying certain laws, but we only dimly understand these laws. Our limited minds cannot grasp the mysterious force that moves the constellations." This is the view of the most creative and intelligent man in twentieth century physics.

Einstein is not alone in this thought process. Dr. Paul Davies, writing in his book *The Mind of God*, writes:

> I belong to the group of scientists who do not subscribe to a conventional religion but nevertheless deny that the universe is a purposeless accident. Through my scientific work I have come to believe more and more strongly that the physical universe is put together with an ingenuity so astonishing that I cannot accept it merely as a brute fact. There must, it seems to me, be a deeper level of explanation. Further, I have come to the point of view that mind—i.e. conscious awareness of the world—is not a meaningless and incidental quirk of nature, but an absolutely fundamental facet of reality.

Dr. Neil deGrasse Tyson, an astrophysicist, writing in his book *Origin of Life on Earth*, more than fifty years after Einstein's death, totally agrees. He writes: "The origin of life on Earth remains locked in murky uncertainty. Our ignorance about life's beginnings stems in large part from the fact that whatever events made inanimate matter come alive occurred billions of years ago and left no definitive traces behind." (page 235)

Another Astrophysicist, Dr. Marcelo Gleiser, puts it this way: "To know the universe is to know ourselves." So as far as astrophysicists and theoretical physicists are concerned, human beings were created directly out of the universe.

Backing this point of view is yet another experimental physicist, Dr. Wenlon He. Dr. He first studied physics in Suzhou, Jiangsu Province, China. At the time of this writing, Dr. Wenlong He worked for Scotland's University of Strathclyde.

Given facts some careful thoughts, Wenlong He made the following observation:

A closed system as the Earth cannot become more organized or remain organized unless acted upon by an external agent. That is the second law of thermodynamics. Since the universe and life on earth are highly ordered, they must be products of an external agent, a Creator. The second fact is that the universe and the earth seem to be specifically designed to support life. This arrangement cannot happen by chance or randomly.

The evidence of "design" I see is the fact that practically, all life on earth depends on energy from the Sun. This energy travels through space as radiation. It comes to earth in a vast spectrum of wavelengths. The shortest wavelengths are the lethal gamma rays. Then come X-rays, ultraviolet rays, visible light, infrared, microwaves, and the longest of all, radio waves. Remarkably, our atmosphere blocks much harmful radiation while allowing other needed radiation to reach the earth's surface. Our atmosphere's special transparency to light, which plants need to produce food, and we need to see, cannot be a coincidence. Even more remarkable is the tiny amount of ultraviolet light that reaches the earth's surface. Ultraviolet radiation is critical. We need a small amount of it on our skin to produce Vitamin D, which is vital for bone health and growth, and evidently for protection from cancer and other diseases. However, too much of this particular radiation causes skin cancer and eye cataracts. In its natural state, the atmosphere allows only a tiny amount of this ultraviolet radiation to reach the earth's surface—and it is the right amount. This is the evidence that a Creator designed the earth to sustain life.

And may I add, we did not arrive on earth by chance.

Dr. Wenlong He's conclusion is in agreement with Greek science which was based on three fundamental assumptions: that behind all natural phenomena there exist orders, that this order is intrinsic and not arbitrary, and that it can be discovered by the human mind. We must employ the human mind, through pure thought, to discover our origin and the nature of our God.

In my judgment, when we consider the excellently woven integrated complexity of a human cell after almost 4 billion years of evolutionary existence, it is difficult to not conclude that a transcendent Creator designed the cell at a stroke. Given the number of elements in the natural world or universe that directly constitutes the human body at the physiologic level, it is fair to conclude that we as human beings were created by and out of the evolving biosphere by a skillful Designer.

Another well accomplished and successful astrophysicist, Dr Bernard Haisch, writing in his book *The God Theory* wrote:

> The origin of the universe is the exact opposite of random. Our lives are the exact opposite of pointless. It is not matter that creates an illusion of consciousness, but consciousness that creates an illusion of matter. The physical universe and the beings that inhabit it are the conscious creation of a God whose purpose is to experience his own magnificence in the living consciousness of his creation. God actualizes his infinite potential through our experience; God lives in the physical universe through us. Our experience is his experience because ultimately we are him, that is, immortal spiritual beings, offspring of God, temporarily living in the realm of matter. It may be audacious to state things so bluntly, but I would be so foolhardy as to just make up such ideas on my own. They are not mine. They are retrieved from the bottom of a vast overlay of religious dogma. They are jewels I have stolen from the Titanic and labeled it "The God Theory." (page 137)

Einstein would often make a distinction between two types of God, which are often confused in discussion about religion. First, there is the personal God, the God that answers prayers, parts the water, fights wars for certain people, and performs miracles. This is the God of the Bible, the God of intervention, the Jewish/Christian God. Then there is the God Albert Einstein truly believed in, the God of Spinoza, the God that created the simple and elegant laws that govern the universe. Like many theoretical physicists as well as astrophysicists of our time,, Einstein believed the orderliness of our universe by a skillful "Designer."

It is this belief, the belief that man did not evolve from lower animals or from a single molecule through evolution, or by pure chance, that put Dr. Albert Einstein above most theoretical physicists, and was the main reason for his departure from mainstream physics.

He once said, "A man who regards his own life and that of his fellow creatures as being meaningless is not merely unhappy… but hardly fit for life."

Just as no one can simultaneously accept a belief in a personal God who makes miracles and is involved in the affairs of man and still claim to be a logical and rational scientist without engaging in magical thinking, so it is equally true that no one can simultaneously accept the premise of Darwin's "theory of evolution" without engaging in fuzzy and magical thinking. Man evolves from man however primitive.

Again, I return to Dr. Bernard Haisch, astrophysicist, on his view about the universe and our place in it in detail. He writes:

The God Theory posits the existence of an infinite, timeless consciousness that, in religious terms, can be called "One God." In principle, this One God is the same for all religions. The vagaries of human nature, history, and culture have, however, transformed this One God into something that unfortunately, varies dramatically from religion to religion. It is worthy of note here that Kabbalah clearly—and wisely—cautions that all descriptions of God are necessarily wrong, because an infinite, timeless consciousness can have no characteristics that can be properly translated into physical terms. Love, light, and bliss come the closest....

Living as a human being in an imperfect world, you certainly experience a reality that seems far from God-like. Yet that may be the whole point. God, as God the transcendent and omnipotent, has all the perfection that is possible, or that could ever be imagined. But "perfection without experience is like a symphony that is never performed, an opera never staged."

The life you experience is a divine exploration, in and through the physical, of the power of infinite creativity. An experience cannot be had without imperfection. Imperfection is absolutely necessary to experience. The problem facing mankind today is that the degree of imperfection has gone far beyond a healthy polarity. This is due primarily to ignorance of what we truly are—immortal spiritual beings—and what the purpose of creation is—God transforming infinite potential into actual experience through us and all other living things. That you remain unaware of your participation in this exploration appears to be necessary part of the experience of creation."

You do have the power to set aside the unhealthy dogmas of both religion and scientism. You can open your mind and use reason and intuition in roughly equal measure to figure out what you truly are. And that will change the world. "Science without religion is lame," wrote Albert Einstein, "and religion without science is blind." Max Planck, one of the founders of quantum mechanics, agrees. "Modern physics impresses us particularly with the truth of the old doctrine which teaches that there are realities existing apart from our sense-perceptions, and that there are problems and conflicts where these realities are of greater value for us than the richest treasures of the world of experience."

Like Einstein and Planck, many of the great scientists of the twentieth century recognized that modern physical science was a "special case" dealing with a "subset of experience." Physical science has obviously proven to be remarkably rich and productive within its material domain, but it would be naïve to assume that "the richest treasures of the world of experience" could substitute for genuine and full knowledge of reality, and much less wisdom.

"In the physics laboratories of today, we acknowledge an enigmatic, but undeniable, relationship between consciousness and the outcome of quantum experiments. In the history of mankind, we acknowledge that the aggregate of direct human experience does not fit within the artificial confines of physical laws. There is simply more to reality than physics, something the majority of humanity seems to know intuitively."

"At the heart of quantum physics is the concept of complementarity, which holds that the simultaneous measurement of wave-like of particle-like properties of matter is impossible and contradictory. This theoretical dilemma is resolved by the conviction that both descriptions are at the same time true, yet incomplete. I propose that a similar, but even higher- level, principle of complementarity exists between reality as a scientific experiment and reality as a spiritual experience. Today, however, rather than seeking a metaphysical principle of complementarity in which scientific experiment and spiritual experience are different perceptions of the same reality, science is attempting to subsume spirit under science. Sometimes this amounts to more than cynical debunking; sometimes it is philosophically florid enough to look profound. In either case, the end result is an attempt to "explain away" any truly spiritual realm."

Science is driven by a spirit of inquiry and methodical investigation and analysis. It is highly successful enterprise for the investigation of the physical world. But to claim that the investigation of the physical world rules out inquiry into anything spiritual is both irrational and dogmatic. To reject evidence simply on the grounds that it cannot yet be measured with instruments in a laboratory is contrary to the scientific spirit of inquiry. It is time to move beyond this fundamentalist science model.

"I think the situation will be radically different in the future. I do not think that this new century will be dominated by merely inanimate high technology. It is my view that exploring and discovering the latent transcendent powers of our own creative consciousness will be more important and of greater value to civilization. Indeed, this will be a kind of circling rather than a wholly new direction in human history. There may be a spiritual substance to science and as well as a scientific substance to spirit, with light as the link. The time has come to reintegrate the two."

The challenge for the institution of modern science is to be true to its fundamental commitment to examine evidence. Scientists must resist the temptation to explain away evidence like near death experiences, simply because they contradict the reductionist paradigm. The analogous challenge for religion is to replace dogma with unfettered search for an experiential truth. Ironically, religion may put itself out of business if it successfully elevates humanity to a level of consciousness that no longer requires a middleman. In my judgment this would be a good thing given the many unspiritual factors and unsubstantiated claims that have influenced organized religion. On the one hand, I think we will practice some form of science forever, provided that science can evolve beyond the constraints of its reductionist ideology. One the other hand, religion must make a huge makeover to stay away from making claims it has no plan to make sense of. Curiosity is, after all, an essential trait of human consciousness. There should be no attempt to control or condition it (Haisch, 2014, The God Theory).

I have included lengthy discussions and quotations from some of the best scientists of our time to draw attention to apologists of Charles Darwin's evolutionary theory that to explain away perceptions because they don't fit in with the reductionist paradigm is not science proper. Darwin's hypothesis on the origin of species and that of evolution lacked scientific foundation. In fact, a thorough analysis and investigation of Darwin's *Origin of Species* clearly showed that Darwin did not address the actual origin of any species including human beings. For some strange reason, Darwin's supporters who are scientists still lay claims that Darwin provided evidence about the origin of species. Darwin attempted to discuss the evolution of species in his own right, but not the origin of species. There is a world of difference between the word evolution and origin. Moreover, for a man to write a huge book on the origins of species without detailing how the origin of species occurred is also problematic. The fact, therefore, must be stressed: Darwin's theory of evolution becomes relevant only, and only after life has already begun. The origin of life on Earth will remain a scientific mystery, as it is beyond discovery in the scientific realm. Maybe, a revision of the Cambrian Explosive Era may provide the answer as to our origin.

This tiny book is not just the isolated views of its author and others cited in it because by the beginning of the 1900s, the majority of scientist in Britain, Europe, and the United States had also rejected large parts of Darwin's "theory of evolution." Most scientists accepted the basic idea of evolution: living things evolve from earlier forms of life. Humans evolved from humans, apes evolved from apes, whales evolved from whales, etc. Like me, they had serious doubts, however, about natural selection as the force that randomly made evolution work.

Now that Darwin's "theory" has been overthrown over the sheer weight of logical simplicity or reason, the question is then asked: Where did we come from and how did we get here? As you know by now, well learned and accomplished physicists and astrophysicist do not believe in the randomness of our universe or the origin of humans from molecules. The Cambrian Era may be the answer to this age-old question. With an open mind I invite to you to read and appreciate the passages that follow. These passages represent my thinking on the origin of the human race.

"Whatever there is of God and goodness in the universe, it must work itself out and express itself through us. We cannot stand aside and let God do it for us."

Albert Einstein, 1940 Einstein Archives

"Our understanding of the most fundamental truths governing our universe is shifting. Soon, we may begin to understand the ultimate laws of nature and to formulate our human estimation of God's equation. When the final equation is constructed, we should be able to use it to solve the wonderful riddle of creation. And perhaps that's why God sent us here in the first place."

Amir D. Acel

Creation: The Cambrian Era
"If, then, it is true that the axiomatic basis of theoretical physics cannot be extracted from experience but must be freely invented, can we ever find the right way? I answer without hesitation that there is, in my opinion, a right way, and that we are capable of finding it. I hold it true that pure thought can grasp reality, as the ancients dream."

(Albert Einstein, 1954)

The Big Bang

Accordingly, mainstream science traces the history of the cosmos back to a Big Bang some fourteen billion years ago, a theory that is correct as far as it goes. "Science deals with the aspect of reality and human experience that lends itself to a particular method of inquiry susceptible to empirical observation, quantification and measurement, repeatability, and inter-subjective verification-meaning, more than one person has to be able to say, "Yes, I saw the same thing. I got the same results." So legitimate scientific investigation is limited to the physical world, including the human body, astronomical bodies, measurable energy, and how structures work. The empirical findings generated in this way form the basis for further experimentation and for generalizations that can be incorporated into the wilder body of scientific knowledge (Haisch, 2014). This is effectively the current paradigm of what constitutes science as we know it. Clearly, this paradigm dos not and cannot exhaust all aspects of reality, particularly the origin of our cosmos and the nature of human existence. As far as the conclusion of the Big Bang goes, there remained many unanswered questions. The questions not addressed over the many centuries by the theory, however, are what caused the Big Bang? Where did the primordial stuff of the explosion originate and when did time begin? If the universe has only one beginning, was there only one Big Bang, or were there many? Is there one universe, or are there many universes? There is no single widely accepted answer to these questions. Some scientists simply say that no one knows or can know the origin of the cosmos and admit that the questions lie beyond the scope of science (7). But whatever the story concerning many universes, we know that our universe or cosmos is the only one that supports life. There will be no one to observe other universes if they exist.

Ironically, however, by casting these questions beyond logic and the scope of science, science, in a sense, admits the possibility that the riddle of the origin of the universe and of man in it requires that we look beyond the laws of science, at least as they are defined today. Of course, this route takes us directly into the realm of creation and religion—an approach that most scientists reject (7).

Because scientific investigation is limited to quantification, measurement, and formulation of empirical data and equations, the origin of sentience beings will never be known through scientific investigation. However, through "pure thought" we may be able to make sense out of our origin. The discussions that follow are based on research and study and my personal reflection as I see the relevance and meaning of life. Man is created in the image of God. "To know the universe is to know ourselves."

"In the beginning was the world, the world was with God, and the world was God." (Genesis 1:1)

The discussion that follows is my interpretation of the events that led to our arrival on Earth based on research, and pure thought, or study as I said earlier. However, I do not pretend to have definitive answers to these ageless concerns. If great minds such as Aristotle, Lao-tzu, Buddha, Socrates, Descartes, Rousseau, Rumi, Goethe, Patanjali, Spinoza, St. Francis, St. Thomas, Shaw, Whitman, Einstein, Haisch (among many others) could not come up with definite and decisive answers to this human puzzle of our origin, then it is clear that I with my very limited training in matters of philosophy, religion, and science will not be able to solve this mystery in this tiny book, or even during my lifetime. The attempt here is to offer my own interpretation of what I have come to appreciate through "logical simplicity" or reason of how we came into existence as physical human beings. The means of our arrival on planet Earth is through Creation via the Cambrian Era. Simply put, life on Earth has resulted from God's direct intervention, converting the constellation of energy or biosphere into early human beings as our ancestors. Let me explain.

The premise of this discussion is that we have always existed before our current appearance of Earth. This being the case, let's consider for a moment the events that occurred and the people that existed prior to our arrival on planet Earth in our generation as human beings. The assumption here is that this world has always existed before our time, and that we were and are spiritual beings. We must wonder about what determined the precise time of our arrival. That is cause and effect. So where were members of this generation when Aristotle, Plato, Hippocrates, Pythagoras, Julius Caesar walked on this planet? What were we doing during the 12th and 13th centuries while the Christian Crusades were taking place? Or even before that, what were we doing in 2500 BCE when the pyramids were being built by our Egyptian brothers? Where were we 1 million years ago when the dinosaurs were roaming the Earth? Obviously, contemplating questions of this nature led me to study a fair amount of science that explained how things came into form. I could not rely on philosophy or religion alone for answers to these inquiries for these two can only speculate. As far as the religion of the Bible is concerned, the human race evolved from the "seed" of Adam and Eve, with no explanation about their race or even the location of the "Garden of Eden." To think about the possibility of a snake speaking Hebrew or any established language to humans in deceiving Eve is itself tiring. However, while I am by no evidence an expert in this area, this is what science tells me about the origin of life.

It is my understanding and conviction that quantum physics regards this fact as scientifically unassailable: that at the tiniest subatomic level, particles themselves don't originate from pre-existing particles. This means that matter originates from something that is formless. Scientists call the formlessness that produces matter "Energy."[Dyer, p-5] This nonmaterial energy produced the particles that became you and me, however primitive, from the beginning of time. I think of this scenario and of our being as a "shift from energy to form" through the Cambrian Explosion or Era.

The Process as I See It

I imagine the tiny speck of human protoplasm that was my very first particle of humanity as being a part of some kind of pull from a constellation of huge energy source that shifted into a fetus, and then into a baby, a toddler, a little boy, and adolescent, a young man, a mature man, an adult, and now a middle-aged man who has continued to breathe God's air for more than fifty years now. This is how you were also brought into this world. All of those shifts were inherent or present in that originating energy that materialized as a microscopic particle and became you and me during a special event—the Cambrian Explosion.

I can't prove this scenario by any scientific method or laboratory experiment, neither do I have any empirical data to justify my thinking. It is beyond my ability to figure out scientifically how such a miraculous unfolding could take place in the formation of who we are as physical human beings. One thing is crystal clear to me: all of this transpired independently of my ability and yours to do much about it other than simply to observe our development. So where did that tiny little microscopic dot that was our first experience as material subject come from?

We need to remember that quantum physics tells us categorically and emphatically that particles do not come from pre-existing particles. If we reduce that original particle to its subatomic status, it is even smaller than chromosomes, atoms, electrons within the atom, and even the sub-subatomic particles scientists called quarts. Scientists have placed a quart the size of my originating point and yours into a particle accelerator revved up to 250,000 mph and have forced it to collide with another quart. And the result? Nothing!! Nothing was there. It appears that nothing exists at the moment of the transition to something. Simply put, we came from nowhere (formless spirit) to now here. All that exists in the world is from pure formless energy—no particles, no big bang, no Adam or Eve. We simply exploded from our formless energy forms into the human race landing where Africa sits today.

Modern physics confirms the metaphysics of Genesis, which tells us that everything came from God (pure formless energy) and it was all good. In the beginning was the world (constellation of formless energy) and the world was God. Similarly, the Tao Te Ching and Buddha tell us that all being originates in nonbeing [Dyer, 2005]. Thus, the question of where we came from is accurately answered by both physics and metaphysics. They both conclude that the human race originated from something that has no form, no boundaries, no beginning, and no substance. In other words, we came from nothing —pure formless energy. We are all spiritual beings having a temporary human experience. This is our essence. This is where we came from—formless energy (God), the universe. This is how I see it. However, one of the most creative men in twenty century physics, Dr. Albert Einstein, put it this way:

A human being is part of the whole called by us universe, a part limited in time and space. We experience ourselves, our thoughts and feelings as something separate from the rest. A kind of optical delusion of consciousness. This delusion is a kind of prison for us, restricting us to our personal desires and to affection for a few persons nearest to us. Our task must be to free ourselves from the prison by widening our circle of compassion to embrace all living creatures and the whole of nature in its beauty. The true value of a human being is determined by the measure and the sense in which they have obtained liberation from self.

So everything in this material world must be like what it came from, including each and every one of us. We are part of this formless energy we called the universe. Forms are condensed energy.

Regarding the origin of the universe and our place in it, a Greek scholar, Origen, put it right. Origen, a Christian scholar, postulated that the human soul exists as part of the universe before it is incarnated into a physical human body and then passes from one body to another through rebirths until it is reunited with God or the universe, after which it no longer takes a physical form. He believed that all souls eventually return to the universe. As you read on you will see similar philosophical thoughts within Buddhism.

As an analogy, our existence in this world is similar to blood being drawn for a diagnostic test. A small syringe of blood provides the medical practitioner or physician with vital information about the entire supply of the person it was drawn from. It can tell the physician the status of that patient's cholesterol levels, electrolytes, or any disease pattern. Why? Because the blood sample is exactly what it came from. This analogy follows exactly our composition. Imagine the free chemical elements in the universe that also constitute our physiology—hydrogen, oxygen, carbon, and nitrogen. All the other elements together contribute less than one percent of the mass of the human body. The elements beyond the big four include small amounts of phosphorus, which ranks as the most important, and is essential to most forms of life, together with still smaller amounts of sulfur, sodium, magnesium, chlorine, calcium, and iron. All of these elements are found in the natural universe, and in the human body. It is logical and reasonable to believe that these elements were condensed by the "Designer" to create the form we call the human body.

It is logical to indicate that since we did not really come from our parents, our religion, or culture, or anything else in this world, it is not necessary or required that we must be like our parents, our surroundings, or society. But since we came from an invisible energy source some people called God, Divine Mind, Yahweh, Yonswa, or universe, then we must be like what we came from. The logical conclusion about our origination is that we came from formlessness or spiritual form, and our true essence is that we are a divine piece of God or universe. We are first, foremost, and always spiritual beings inextricably connected to our source of being [Dyer, 2005). We did not evolve from lower animals or from a single lifeless molecule! As suggested by Einstein, "pure thought" will guide our way in discovering our true origin. Personally, I view the Cambrian Explosion as the "origin of species."

Robert Burns summed this experience up in his famous poem, "New Year's Day," written in 1791.

The voice of Nature loudly cries,
And many a message from the Skies,
That something in us never dies

That which is formless cannot be destroyed. That which is formless energy cannot die. The formless aspect of all being exists in eternity, impervious to beginnings or endings. The real truth seems to be that our essence is eternal, and that it is only the physical body that appears to come and go in a cycle of birth and death. What we call birth and death are actually as inseparable as two sides of a coin, or daytime and nighttime [12].

We are certainly like God, or the universe, and we have the freedom to make choices. Some of our choices cause our spiritual link to source to become contaminated and rusty. One of those very serious choices is believing that the expression of God through our physical self is an endpoint, or ultimate, rather than an opportunity to choose how to express this gift. In this manner, we edge God out and create an ego-driven life. Those cultures that do not follow the ego-driven life we called primitive people or uncivilized individuals today in our system of acquiring. The great lesson in this philosophical journey is to recognize our primary identity as spiritual being who is eternal and therefore imperious to both birth and death.

Our physical self is an expression in the form of the energy of our spiritual essence; our real self is the loving observer of our sensory experiences. In order to fully harmonize with that essential nature, we must be dedicated to expressing its energy and be fully aware of the sacred choice we continue to make in this world. For some, that will mean becoming more like God while temporarily housed in their bodies as the Buddhists do. For others, it will be creating godlike expressions of beauty, purpose, and wisdom in form as indigenous people do.

The human voyage in bodily form is barely a parenthesis in the eternity of our real self. When the parenthesis closes, we are fully reimmersed in Spirit sans the materialized self.[3] We cannot articulate this process from the scientific point of view by providing empirical evidence, but one of the greatest poets of all time, T.S. Eliot, summed it up in his famous poem "Little Gidding":

> We shall not cease from exploration
> And the end of all our exploring
> Will be to arrive where we started
> And know the place for the first time

Neither philosophy nor religion will show us the way from whence we came. Only pure thought can grasp this reality as the ancients dreamed, apparently during the Cambrian Era.

Understanding the answers to "Where did we come from" involves, more than anything else, attempting to live from a perspective that is in rapport with our original nature. We must become more like the spiritual nature of our origin. By recognizing the expression of Divine consciousness that is our physical being, we in turn make the choice of how to express that Divine spirit. It is needless to believe that our physical being is all there is to our existence. So often our physical world doesn't seem to be so spiritual, in spite of our having originated from spiritual essence. Our life does not end in our physical death. Henry Wadsworth Longfellow expressed this dilemma in one of his famous poems "A Psalm of Life":

Life is real! Life is earnest!
And the grave is not its goal;
Dust thou art, to dust returnest,
Was not spoken of the soul (energy).

The poet speaks of your life and mine as something beyond physical death, which he describes as dust. We are all something other than what we identify with our senses. There is no such thing as a grave for our essential essence—our spirit—but we may disregard and thus, lose touch with it. In fact, that's a pretty common situation for all of us during different periods of our lives when we make the choice to put our physical self in charge. But we also experience dreams in which our spiritual self travels to places and does things with other people in different places while our physical self is sleeping somewhere else!! Our true essence is spirit.

Creativity or pure thought can grasp reality. To write about nonbeing as the place we originate from requires me to imaginatively speculate on what the spiritual world of nonbeing really is. The way to do this is to imagine a Divine consciousness or Constellation of energy who's in the business of manifesting form out of nothingness. Imagining a creation without a Creator is equivalent to imagining a watch without a watchmaker! How can this be? Observing nature or creation every day, I can clearly see that it is from seeds that blossoms come, from blossoms come fruit, and from small acorns come giant oak trees. [15] We all came from a spiritual world to this physical world—begotten, not made.

Nonbeing is a deliriously paradoxical state to contemplate because I know in my heart that it surely exists, yet I only have my limited human mind with which to do the contemplating. Also I know that that field from which all things originate and to which all things return has a feel to it. Based on research, study, and ruminations of some of the most revered human beings who have walked on this Earth, creation is not an act of violence; it is a pleasurable, joyful act. All of creation emerges from a silent void, as does every sound. Every bit of light comes from nothingness; every thought emerges from non-thought. There is a Zen proverb that reminds us that it's the silence between the notes that makes the music. Without silence to interrupt the sounds, there can be no music; it would be only one long, continuous tone. But even the long tone originated in the void. [15]

Nothingness is equivalent to the expression of zero, mathematically. It can't be divided, it has no empirical value, and if we multiply anything by it, we get a sum of nothing. Yet, without the indivisible zero, mathematics itself will be impossible. Before we came into this material world, our essence was nothing. We had no things encumbering us, no rules, no duties, no money, no car, no fear, no parents, no wife or husband, no children, absolutely nothing at all. When we are no longer animated by divine consciousness, an experience called death, we return from where we came, with nothing. No money, no clothes, no children, no wife or husband, no duties, or membership into anything.

Almost every great spiritual master and enlightened philosopher tells us to find God in emptiness, to imagine our origin in nothingness, and to hear God speak to us in silence, as all these are nothing but spiritual connectedness. So one of the answers to the question of where we came from is: Nowhere, and with nothing.

We must make the effort to find our way to that peaceful nothingness while we are still in our physical body. We can empty our pockets or purse, but we especially need to empty our mind and relish the joy of living in our physical world while simultaneously experiencing the bliss of nothingness. This is, indeed, our origin, just as it is assuredly our ultimate destination as well. Albert Einstein, the world best known theoretical physicist once noted that "everything is emptiness, and form is condensed emptiness."

There is another philosophical thought about the origin of human beings or "sentient beings," as far as the natives of Konobo tribesmen are concerned. In the earliest oral history told to us over the many years, there is an illusion to a story of human origin and evolution, which is recounted in many of the subsequent oral talks in our villages in Konobo District. The story unfolds in the following manner.

The Konobo traditionists' cosmos consists of three realms of existence—the desire realm, the form realm, and the formless realm, the last being progressively subtler states of existence. The desire realm is characterized by the experience of sensual desires and pain. This is the realm that we human beings and animals inhibit. In contrast, the form realm is free from any manifest experience of pain and is permeated principally by an experience of bliss. "Beings in this realm possess bodies composed of light." Finally, the formless realm utterly transcends all physical sensation. Existence in this realm is permeated by an abiding state of perfect equanimity, and the beings in this realm are entirely free from material embodiment. They exist only on an immaterial, mental plane. Beings in the highest states of the desire realm and those in both the form and formless realms are described as celestial beings. They are not permanent, heavenly states to which we should aspire. I suppose this was our form of existence shortly before the Cambrian explosion!!

The origin and evolution of human life on Earth is understood in terms of the "descent" of some of these celestial beings who have exhausted their positive energy and action, which provided them with the cause and conditions to remain in these higher realms. There was no original sin that precipitated the fall; it's simply the nature of impermanent existence, cause and effect, that causes a being to change states or "to die." When these beings first experienced their "fall" and were born on Earth during the Cambrian Era they still possessed the vestiges of their previous glories. These humans of the first era were thought to have godlike qualities. They are said to have come into being through "spontaneous birth." They had attractive physiques, their bodies had halos, they had certain supernatural powers like flying, and they subsisted on the nourishment of inner contemplation. They were also thought to be free from many of the features that serve as the basis for discriminating identity, such as race or caste. These were the first human beings placed on Earth. It is fair to say they lived in what is Africa today because modern genetics placed the origin of the human race in Africa. So the evolution of humans did not begin with a man and a woman. It probably began with these celestial beings, maybe two hundred thousands of them at once in what is Africa today. (pictures/photos early people appear here)

Over time, it is said, these humans began to lose these qualities. As they took nourishment from material food, their bodies assumed coarser corporeality and thus gave rise to a great diversity of physical appearances. This diversity, in turn, led to feelings of discrimination, especially animosity toward those who appeared different, and attachment toward those who were similar, resulting in the emergence of the whole host of gross negative emotions such as fighting and possession. Furthermore, the dependence on material food led to the need for disposal of waste from the body. The story continues with a detailed account of the genesis of the entire range of negative human actions, such as killing and sexual misconduct. During this time, no particular female belonged to a particular male, and no children belonged to a particular "family." The universe, or God, transformed himself into the human race. (See the plates on early human beings as they descended from the celestial forms during the Cambrian Era.)

Central to this account of human evolution is the Konobo traditionists' hypothesis of the four types of birth. In this view, sentient beings can come into being as (1) womb-born, such as we humans; (2) egg-born, such as the birds and many reptiles; (3) heat and moisture-born, such the numerous type of insects; and (4) and spontaneously-born, such as the earlier celestial beings in the form and formless realms. As to the question of diversity of life, Kwee Yeali philosopher expressed a common viewpoint when they stated " it is from the mind that the world of sentience arises. So too from the mind the diverse habitats of being arise."

It should be noted that because of the indivisibility of consciousness and energy, there is a profound intimate correlation between the elements within our bodies and the natural elements in the universe, examples being hydrogen, oxygen, nitrogen, carbon, potassium, iron, etc., as discussed earlier. This subtle can be discerned by individuals who have gained a certain level of spiritual realization or who have naturally higher level of perception. In fact, according to Kwee Yeali thought, there is an understanding that our bodies represent microcosmic images of the greater macrocosmic world. Simply put, our bodies are condensed energy, as evidenced by the natural elements that are part of the human physiology. These are the philosophical views of Konobo tribesmen's thoughts which directly correlate to the philosophical views of natural theologians. Through oral stories, the elders of Konobo-speaking Africans taught their young people that Yonswa (God) has no beginning and that "Kwee," the spiritual leader of the clan, is God incarnate. The majority of Konob-speaking Liberians still do not accept the biblical concept of Creation. They subscribe to the teachings of the Konobo elders which is similar to the Cambrian Explosion Era.

In support of the philosophy of the Konobo people of Africa the facts may be stated: Complex biological systems were never put together by the accumulation of random mutations through natural selection as some scientist suggest. There is no fossil history of single-celled organisms changing forms step by step into complex plants and animals. All major groups of animals, including man, appear suddenly in the rocks of the Cambrian Era, and no next groups appear thereafter with different fossil remains. According to paleontologists and geneticists, the Cambrian Explosion took place in Africa, the cradle of the human race.

Please view evolutionary pictures or plates of our ancestors in their most primitive forms. I believe these were our first forms of existence as sentience beings, emerging from our formless energy forms. This formless energy transformed into matter which we are.

"The nature of God is a Circle of which the Center is everywhere and the Circumference is nowhere. This universe, which is the same for all, has not been made by a man, or a god, but it has always been, is, and will be forever, kindling itself by regular measures, and going out by regular measures."

Empedocles

Conclusion

My conclusion is that everything is energy; it's all vibration at a variety of frequencies. The universe is nothing but a constellation of energy of which we are a part. This constellation of energy (universe) transformed into matter (the human race via the Cambrian Explosion). The faster the vibration, the closer one is to Spirit and understanding where we came from. The pencil I hold in my hand as I compose this paragraph appears to be solid, yet a glance at it through a powerful electron microscope shows that it is actually a field of moving particles, with mostly empty space between those particles. The vibrational makeup of my pencil is energy that is slow enough to appear solid to my eyes, which can only perceive objects that fit within certain frequency. Try this experiment when you have a powerful microscope. We are this constellation of energy currently in our physical bodily form [15]. At times this spiritual part of our physical body leaves the body when you dream and go places or do things with others— events you clearly remember when you are awake! Certainly, we did not evolve from apes or a single organic molecule. We came from spirit form—pure energy. Certainly, we return in spirit form when this life is no longer animated by Divine consciousness, when death takes over.

Everything in this world can't be quantified but everything in this world must follow some logical simplicity. No amount of science will provide with clarity the origin of man. For example, paleontology and archeology can only trace our evolution from our former Homo sapiens status to our modern status through the study of fossils. Philosophy can only speculate about the origin, while religion can only make claims about our origin. We may never know with certainty how we evolved from our spiritual essence to our current human form on an empirical basis, but at least we can reason how this might have taken place. However, this does not mean we can engage in fuzzy thinking about our evolution. Maybe the branch of science called "Thermodynamics" can give us a clue of our transformation from pure energy (nothingness) to our current form (condensed energy—Conservation of Mass/Energy).

By definition, thermodynamics is the science of energy transformation. Here, we have to put things into context. So instead of definition, we deal with application. Let's look at Einstein's $E=mc^2$. Most people have heard of or seen this equation, but many people may not know what it really means. This equation is read "E equals mc squared." *E* in this equation stands for energy, *m* stands for mass, and *c* is the speed of light. Squaring a number means multiplying it by itself.

This simple-looking equation is a mathematical way of stating a very big and impressive idea. It says that matter and energy are actually different forms of the same thing! Matter can be converted into energy and energy can be converted into matter. In a way it is logical to think of all matter in our universe as frozen or condensed energy. A human being is matter, a fact based on science.

For example, look at a match, a pencil, or even your little finger. Whatever piece of matter you choose to look at is composed of billions of tiny atoms. Put any of your choice under a powerful electron microscope. Each atom is vibrating with energy. Energy holds the atoms together and in place. Energy connects the atom with one another to form molecules. And energy even holds the subatomic particles together to form atoms themselves. Each atom is a whirling, pulsing clump of energy we call matter.

Einstein's equation tells us that there is a tremendous amount of energy locked up even in the tiniest bits of matter. The equation tells us how to figure out how much energy (E) there is in a piece of matter. We must multiply the amount of matter (m) by the speed of light squared. The speed of light, of course, is a very large number. Squaring the speed of light gives us an enormous number. If we could somehow release all the energy in a single gram of matter, we would have a tremendous amount of energy. One gram of matter is equal to more than 21 billion calories of energy. This is the amount of energy released by burning 2,700 metric tons (3,000 tons) of coal! (Geordi, p-34)

Einstein's equation also tells us that if we gave an object more energy, we could increase its mass. To do this, we would have to add tremendous amounts of extra energy. You may be wondering if these events are actually possible. Is it really possible to change a piece of matter into energy? Could we somehow turn matter into energy? The answer is YES. Both transformations have been done here on Earth during the last fifty years.

Scientists have changed energy into mass in giant particle accelerators that are used in modern physics research. A particle accelerator is a very huge machine that boosts the speed of tiny atomic particles almost to the speed of light. These accelerators are circular tubes, sometimes miles around. Atomic particles are held in these tubes and sped up with electromagnetic force. Powerful magnets keep the particles from flying out through the walls of the tube. The atomic particles, such as protons or electrons, whirl around the ring of the tubes at enormous speeds. These tubes have no air in them, and so the particles have no atoms to collide against. Each time they zoom around the ring they are given another boost of power from an electromagnetic field. They go a little faster with each boost. Eventually, after traveling around the tube millions of times and receiving millions of energy boosts, the particles are traveling almost at the speed of light. Because of their speed, they have tremendous energy. The beam of particles is then aimed at a target. Of course, the atomic particles are too fast and much too small to see. So physicists find out about them by examining evidence they leave behind when they hit the target. Researchers can keep a careful photographic record of what happens to the target as the particles hit.

Now let me give an everyday example. Imagine that you arrive at the scene of an accident. A vehicle has crashed into a wall, but the vehicle has been towed away. Somehow you know the vehicle that hit the wall was moving at fifty miles per hour. It should be easy for you to tell whether the vehicle was a bicycle, a small car, or a large truck by looking at how much damage has been done. Certainly, a truck will cause much more destruction than the bicycle.

Researchers and scientists study the results of the collision in their accelerators in a similar way. They can't see the speeding particles, but they know how much energy the particles were given as they were accelerated. These scientists study the evidence the collisions of particles left behind. For example, from studying photographs of very fast protons hitting their targets, scientist can tell that these particles have more mass than ordinary protons. The energy that they received as they accelerated to tremendous speeds has been converted into extra mass. So particle accelerators have actually managed to turn tremendous amounts of energy into tiny bits of matter.

Can matter turn into energy? Yes, indeed. That has been done as well, in nuclear reactors and nuclear bombs. A nuclear bomb has about 10 kilograms or 20 pounds of uranium or plutonium metal for fuel. In a nuclear reaction, for example, the atom splits apart, forming atoms of other elements. When an atom splits, some of its mass is released as energy. When a nuclear bomb explodes, about 1/1000 of its nuclear fuel is converted from matter into energy. So a small atomic bomb explodes with the force of only 10 grams (0.4 ounces) of matter! But the energy in that small amount of matter creates the enormous explosion and blast of the bomb.

The same conversion of matter into energy takes place in a nuclear reactor. In nuclear reactor, the conversion from matter to energy is carefully controlled to avoid an explosion. The radioactive fuel in a nuclear reactor is extremely dangerous to handle. But suppose you could weigh the fuel that's put into a reactor, allow it to run for a year, and then weigh the fuel again. You would find that a small fraction of the weight has disappeared. Some of the uranium in the reactor would have been converted to the heat that is then used to generate electricity.

Our Sun also changes mass into energy. The Sun is a giant nuclear reactor, constantly transforming part of its mass into energy. We receive some of the Sun's energy here on Earth as sunlight.

Another simple example is a match. You can see mass transformed into energy right before your very eyes. Just light a match. Tiny quantities of mass are converted to energy even in ordinary chemical reactions. When a match burns, a minute amount of the match's mass is released as energy. We see that energy as light and heat. When you touch the light from the match you will definitely feel the heat.

Einstein's equation shows that mass can be changed into energy and that energy can be changed into mass. We can no longer say that the total amount of mass still remains the same in every reaction. Some mass is lost, becoming energy. Also, we can no longer say that the total amount of energy in any reaction always remains the same. Some energy may be converted into mass! The law of conservation of matter is not true. Matter and energy are equivalent. Matter is simply energy in a different form, and energy is matter still in a different form. For this reason, and to be exact, scientists now use the law of conservation of Mass/Energy. It states that the total amount of mass and energy in any reaction must remain the same. Mass may be converted into energy or energy may be converted to mass, but no mass or energy can be created or destroyed. [Fleisher, p 45]

All those important changes came from Einstein's simple equation: E= mc². It has finally given us a whole new way of looking at our universe and ourselves. Einstein came very close to understanding the nature of God and of the universe, and our place in it. But there still remains uncertainty among those who earnestly try to understand our universe. Werner Heisenberg believes that this uncertainty will remain as long as we live in this world.

The instruments of science have become better and better over the past four hundred years. Scientific measurements have become more and more accurate. In the early 1900s, it seemed as if anything scientists wanted to know could be measured It was just a matter of making a measuring device that was accurate enough and then using it to look at whatever was there to measured. At that time, what scientists most wanted to look at were atomic particles. So here Heisenberg's "Uncertainty Principle" is given as an example that we will never be certain about everything in our universe. The law can be stated like this: It is impossible to measure atomic particles without disturbing the particles. Therefore, it is never possible to know everything there is to know about these particles.

The Heisenberg Uncertainty Principle is a very important idea for both scientists and many other thinkers. It tells us that no matter how accurate or careful our scientific measurements become, we can never know everything about the universe. If nothing else, there will always be uncertainty about the actions of atomic particles. We can never be sure what they are doing at any given time. The best that physicists can do is to determine where an electron is most likely to be at any time, and how fast it is most likely to be moving. [Fleisher, p 44]

Heisenberg's Uncertainty Principle is a law that applies to the actions of atomic particles. But the idea is an important one for other sciences too. All scientists need to remember that the process of observing something often has an effect on the thing they are observing. Therefore, it may be impossible to observe any event perfectly natural, undisturbed conditions.

For example, biologists often look through a microscope at tiny creatures in a drop of water. But the very act of looking at those creatures can change the creatures' behavior. They have been removed from their environment and placed in a single drop of water on a glass slide. A microscope focuses extra light on the subjects being observed so that they can be seen clearly. The light also raises the temperature of the water. All these changes can affect the actions of the creatures the biologist is observing.

Another example is given about the nature of forest animals. Suppose we wanted to study the activities of forest animals without disturbing them or without the interference of human beings. We could set up an automatic camera in the woods and then leave. Even so, the strange appearance of the camera and its whirling sound might affect the animals' behavior in an unexpected ways. Perhaps the scent of the camera of the humans who set it up might also affect what the animals do. No matter how careful we were, we could never be certain that the animals we photographed were behaving in a complete natural way.

Heisenberg's uncertainty principle tells us it is impossible for science to understand the universe completely. No matter how carefully we experiment, no matter how accurate our scientific instruments are, some things in the universe will always be hidden from us. In an effort to find the laws that tell us how our universe works, scientists will continue to experiment. But in a sense, Heisenberg's uncertainty principle may be the last law of the universe. [Fleisher, p43] No matter how wonderful human brains are, or may be, we can't know everything about our universe until our exploring ends with death and we really discover where we came from in spirit form.

Scientists are still learning about stars and the planets, the atom, and the miracles of life. There are still more laws to discover and many more mysteries to solve. But Heisenberg reminds us that we can never be certain about everything in our universe.

Let me end this discussion with a quote from the most creative man in 20th century physics, a man who wanted to "know God's thoughts" —Albert Einstein.

"The most beautiful experience we can have is the mysterious. It is the fundamental emotion that stands at the cradle of true arts and true science. Whoever does not know it and can no longer wonder, nor marvel, is as good as a dead person, and his eyes are dimmed. It was the experience of mystery—even if mixed with fear—that engendered religion. A knowledge of the existence of something we cannot penetrate, our perceptions of the profoundest reason and the most radiant beauty, which only in their primitive forms are accessible to our minds: it is this knowledge and this emotion that constitute true religiosity."

In this sense, and only this sense, all creative thinkers are deeply religious men and women. We must be satisfied with the mystery of life's eternity, and with a knowledge, a sense, of the marvelous structure of existence—as well as the humble attempt we make to understand even a tiny portion of the Reason that manifests itself in nature. Priests and pastors and all those who claim to be holy and knowing can only speculate. I still strongly believe that the human race evolved out of the Cambrian Explosion or Era, irrespective of all the physical principles I have quoted.

Albert Einstein tried to understand the origin and geometry of the universe and of our place in it as human beings. He tried to "know God's thoughts," knowing that "God does not play dice." But he also understood that his efforts had been limited by the unavailability of mathematical methods. In developing "Special Relativity," Einstein used the mathematics of Lorentz and Minkowski. For the theory of General Relativity— which gave us a window into God's creation—he successively used the mathematics of Ricci and Levi-Civita and that of Riemann. But here, Albert Einstein had to stop. He had come a long way toward discovering God's Equation, but he could go no further. For us on this planet, "Mathematicians will attempt to develop the tools, theoretical physicists will apply them, astronomers will verify the theories and provide data, and cosmologists will generate the big picture of the origin of our universe and of our place in it" (J.S. Rigden, 2005). But let me add, we will never know based on empirical formulae the origin of species, including man. We must accept the Cambrian Explosion as the actual origin of the human race. Remember T.S. Eliot's "Little Gidding."

> "Einstein's theories to explain the links between Relativity and cosmological constant give him the apparent role as God's mouthpiece, revealing the most fundamental truths about our larger environment, truths scientists are now just confirming."
>
> (J.S. Rigden, 2005).

Jesus Christ was to the religious world in explaining the nature of God as Albert Einstein was to the scientific world in explaining the nature of God, God's thoughts, the universe, and our place in it.

Charles Darwin does not qualify in any of these categories, neither did he possess the credentials needed to do so. The scientific world must disregard Darwin's "evolutionary theory" on the origin of species and continue the quest to find the origin of our universe and all the species in it through pure thoughts guided by Reason and science. His contribution to the evolution of man, as **being** *from the cradle of Africa is, of course remarkable.*

The origin of life, in my judgment, certainly has resulted from God's direct intervention, converting formless energy to the human race;

The origin of life is beyond discovery in the scientific realm as there is no convincing scientific explanation describing the origin of life to date;

God is imminent in creation, upholding the natural laws which physics attempts to discover;

God's sustaining creative presence undergirds all of life's

history from the beginning to the present, and that;

The theory of evolution as proposed by Charles Darwin and supported by some scientists does not produce any evidence or explanation to the question of the origin of life of species including man at all, and the evolutionary theory becomes relevant only, and only after life has already begun.

I have attempted to make my contribution to this human dilemma on the origin of the human race and to finally close the chapter on Charles Darwin's Origin of Species. I hope the evidence produced in this tiny book and the effort made by all contributors will be perceived from the perspective of reason and pure thought.

Important Scientists

In concluding this book I have included here some bibliographies of a number of very important physicists, mathematicians, and astronomers with particular emphasis on the earlier scientists and the key 20th century theoretical physicists involved in cosmology (quantum theory, relativity, and astrophysics).

These scientists are not arranged in alphabetical order.

Nicolaus Copernicus (1473-1543)
Laid the foundation for modern astronomy.
Nicolaus Copernicus was a Polish astronomer who put forth the theory that the Sun is at rest near the center of the Universe, and that the Earth, spinning on its axis once daily, revolves annually around the Sun. This is called the heliocentric, or Sun-centered, system or theory.

Nicolaus Copernicus is the Latin version of his name, which he chose later in life as was the custom among scientists of his day. His original name may have been Mikolaj Kopernik.

Galileo Galilei (1564-1642)
Galileo Galilei was a Tuscan (Italian) astronomer, physicist, math-

ematician, inventor, and philosopher. He was born in Pisa, and was the oldest of six children in his family. When he was a young man his father sent him to study medicine at the University of Pisa, but Galileo studied mathematics instead. He later became professor and chair of mathematics at the University. He helped to mathematically describe ballistics and the force of friction as it relates to motion. Later, Galileo became interested in optics and astronomy, and in 1609 he built his first telescope and began making observations. Galileo's observations have confirmed Copernicus' model of a heliocentric solar system where planets orbit around the Sun, not around the Earth as the Catholic Church thought.

Blaise Pascal (1623-1662)
Blaise Pascal was a French mathematician, physicist, and religious philosopher, who laid the foundation for the modern theory of Probabilities with the invention of the Pascaline, an early calculator. He was the third of four children, and only son to Etienne and Antoinette.

Christiaan Huygens (1629-1695)
Christiaan Huygens was born to a wealthy Dutch family in the Hague. Educated in science and mathematics, he was one of many physicists to be intrigued by the nature of light. In 1678, Huygens proposed his wave theory of light, which was contrary to the particle theory supported by Newton. One of his greatest contributions to physics was his study of the Pendulum and its applications to timekeeping. With his homemade telescope, he

discovered Saturn's largest moon and more clearly distinguished the shape of Saturn's rings, which were first observed by Galileo.

Sir Isaac Newton (1643-1727)

Isaac Newton was an English philosopher, mathematician, astronomer, physicist, and scientist. Sir Isaac Newton was most famous for his law of gravitation, and was instrumental in the scientific revolution of the 17th century. With discoveries in optics, motion, and mathematics, Newton invented calculus and developed the principles of modern physics. In 1687, he published his most acclaimed work: *Philosophiae Naturalis Principia Mathematica* (Mathematical Principles of Natural Philosophy). He was the only son of a prosperous local farmer, also named Isaac Newton.

Antoine-Laurent Lavoisier (1743-1794)

Lavoisier was a French chemist, who developed a theory of combustion, established a system for naming chemical compounds, and contributed to the law of conservation of matter. He first studied law, but he loved science and displayed a great talent and energy for research. As a social activist, he studied ways to improve French agriculture, water quality, public education, and welfare.

Joseph-Louis Gay-Lussac (1778-1850)

Gay-Lussac was a French physicist and chemist who grew up during the French Revolution. He used balloon excursions to observe magnetism and air composition at varying altitudes. His prominent work was done in the study of gases.

Michael Faraday (1791-1867)

Faraday was a British chemist and physicist who contributed significantly to the study of electromagnetism and electrochemistry.

In 1831, Faraday discovered electromagnetic induction, the principle behind the electric transformer and generator. He was partly responsible for coining many familiar words including *electrode*, *cathode*, and *ion*. Because his family was not well off, Faraday received only a basic formal education. He advanced through apprenticed work to a local bookbinder.

James Clerk Maxwell (1831-1879)

Maxwell was one of the greatest scientists who have ever lived. To him we owe the most significant discovery of our age—the theory of electromagnetism. He is rightly acclaimed as the father of mathematics, astronomy, and engineering.

Albert Einstein once said of Maxwell: "The special theory of relativity owes its origins to Maxwell's equations of the electromagnetic field. Since Maxwell's time," he continued, "physical reality has been thought of as represented by continuous fields, and not capable of mechanical interpretation. This change in the conception of reality is the most profound and most fruitful that physics has experienced since Newton."

James Clerk Maxwell became the first professor of Experimental Physics at Cambridge University where he directed the newly established Cavendish Laboratory.

Max Planck (1858-1947)

Karl Ernst Ludwig Marx Planck, better known as Max, was a

German theoretical physicist considered to be the initial founder of quantum theory and one of the most important physicists of the 20th century. Around the turn of the century, he realized that light and electromagnetic waves were emitted in discrete packets of energy that he called "quanta" which could only take on certain discrete values (multiples of a certain constant) which now bears the name "the Planck constant." This is generally regarded as the first essential stepping stone in the development of quantum theory, which has revolutionized the way we see and understand the sub-atomic world.

Ernest Rutherford (1871-1937)

A consummate experimentalist, Ernest Rutherford was responsible for a remarkable series of discoveries in the fields of radioactivity and nuclear physics. He discovered alpha and beta rays, set forth the laws of radioactive decay, and identified alpha particles as helium nuclei. Most important, he postulated the nuclear structure of the atom.

Guglielmo Marconi (1874-1937)

Marconi was an inventor, physicist, and scientist. Through his experimental wireless telegraphy, he developed the first effective system of radio communication. Marconi founded the London-based Marconi Telegraph Company in 1899. Though his original transmission traveled a mere mile and a half, in 1901, he sent and received the first wireless message across the Atlantic Ocean, from Cornwall, England, to a military base in Newfoundland.

Albert Einstein (1879-1955)

At the start of his scientific work, Einstein realized the inadequacies of Newtonian mechanics and his special theory of relativity stemmed from an attempt to reconcile the laws of mechanics with the laws of the electromagnetic field. He dealt with classical problems of statistical mechanics and problems in which they were merged with quantum theory: this lead to an explanation of the Brownian movement of molecules. He investigated the thermal properties of light with a low radiation density and his observation laid the foundation of the photon theory of light.

In his early days in Berlin, Einstein postulated that the correct interpretation of the special theory of relativity must also furnish a theory of gravitation, and in 1916 he published his paper on the general theory of relativity. During this time he also contributed to the problems of the theory of radiation and statistical mechanics.

Einstein's important work included the following: Special Theory of Relativity (1905), General Theory of Relativity (1916), Investigations on Theory of Brownian Movement (1926), and the Evolution of Physics (1938).

Albert Einstein received honorary degrees in science, medicine, and philosophy from many European and American Universities. HE WAS ONE OF THE THEORETICAL PHYSICISTS WHO KNEW AND BELIEVED IN GOD.

Arthur Eddington (1882-1944)

Sir Arthur Eddington was a prominent English astrophysicist of the early 20th century. He is perhaps best known for his observational confirmation of Einstein's general theory of relativity and the bending of light due to gravity. His early adoption and popular expositions of relativity were instrumental in gaining publicity for the theory and disseminating its ideas to the English-speaking world. Also, he helped develop the first true understanding of stellar processes and the internal structure of stars, and he established the Eddington limit which dictates the natural limit to the luminosity of stars.

Niels Bohr (1885-1962)

Niels Bohr was a Nobel Prize-winning physicist and humanitarian whose revolutionary theories on atomic structures helped shape research worldwide.

Bohr's research led him to theorize in a series of articles that atoms give off electromagnetic radiation as a result of electrons jumping to different levels, departing from Rutherford's views on the subject matter. He worked with Werner Heisenberg and other scientists on a new quantum mechanics principle connected to Bohr's concept of complementarity. The concept asserted that physical properties on an atomic level would be viewed differently depending on experimental parameters, hence explaining why light could be seen as both a particle and a wave, though never both at the same time.

Alexander Friedmann (1888-1925)

Friedmann was a Russian cosmologist and mathematician, who helped develop models that explained the development of the universe. In particular, his solutions to Einstein's field equations provided early evidence of an expanding universe, and the theoretical underpinnings for both the big bang and the steady state models of the universe. The expansion of the universe was finally corroborated several years later by Edwin Hubble's observations in 1929.

Edwin Hubble (1889-1953)

Hubble was an American astronomer who demonstrated the existence of other galaxies, as well as his influential work on astrophysics, and his subsequent namesake, the Hubble Space Telescope. He made huge impact on astronomy and science in general by demonstrating that other galaxies exist besides our own Milky Way galaxy, which changed our view of the universe. The Hubble Space Telescope has provided valuable research data and images since it was carried into orbit in 1990, leading to many breakthroughs in the field of astrophysics.

Werner Heisenberg (1901-1976)

Werner Heisenberg was a German theoretical physicist who made foundational contributions to quantum theory. He is best known for the development of the matrix mechanics formulation of quantum mechanics in 1925 and for asserting the Uncertainty Principle in 1926, although he also made important contributions to nuclear physics, quantum field theory, and particle physics. He received his Nobel Prize in physics in 1932 for the creation of quantum mechanics.

J. Robert Oppenheimer (1904-1967)

Oppenheimer was the director of the Manhattan Project that created the atomic bomb. Oppenheimer's struggle after the war with the morality of building such massive destructive weapon epitomized the moral dilemma that faced scientists who worked to create the atomic and hydrogen bombs.

The decision to focus on these contributors in particular at the expense of many, many others is purely my own and in the interest of expediency. A detailed inclusion of all scientists known to man is certainly beyond the scope of this tiny book.

Glossary

- Blended Inheritance: a theory of inheritance based on the idea that traits passed down by parents are mixed or blended together in their offspring.
- Catastrophism: the theory that geological changes during Earth's history happened abruptly and were caused by catastrophic events such as fire or floods.
- Creationism: the belief that the universe and all living beings were created individually by a divine power. Most Christian creationists accept the account of creation presented in the book of Genesis in the Bible.
- Descent with Modification: a phrase used by Charles Darwin to describe gradual changes in organisms that produced new species. The word "evolution" gradually replaced this term.
- Eugenics: a movement dedicated to the idea that human society could be improved by encouraging "fit" people to reproduce and preventing or discouraging "unfit" from having children.
- Fossils: remains of ancient animals and plants that have been preserved in rock or soil.

- Gemmules: units of inheritance that Darwin believed were produced by individual body cells. According to the theory of pangenesis, gemmules are sent to the reproduction organs and passed on to descendants.
- Genes: the basic units of heredity. Genes are sections of deoxyribonucleic acid (DNA) that are carried on the chromosomes within the nucleus of a cell.
- Germ Plasm: in August Weismann's theory of heredity, a material contained in chromosomes. Germ Plasm allows genetic traits to be passed from parents to offspring.
- Intelligent Design, (ID): a belief based on the idea that the complex features of the natural world must have been created by an intelligent designer.
- Natural Selection: the process by which organisms with traits favorable to their environment survive and reproduce. Those with unfavorable traits gradually die out.
- Natural Theology: a belief that the evidence of design and function in the universe prove the existence of God. William Paley (1743-1805) advocated this view in the early 1800s.
- Neo-Darwinian Synthesis: the theory of evolution that combines Darwin's basic ideas about natural selection with modern genetics. Almost all biologists and many other scientists in the twentieth century accepted this idea.
- Paleontology: the science that studies the evidence of ancient life. It includes the study of fossils, comparative anatomy, geology, and other sciences.
- Pangenesis: Darwin's theory of inheritance, based on the

belief that each body part produces specific genetic information that is sent to the reproductive organs.
- Sexual Reproduction: a form of reproduction in which male sperm cells and female egg cells combine to create a new and unique organism. In forms of asexual reproduction such as cloning, the new organism is an exact copy of the original.
- Social Darwinism: a general term used to apply biology to social and economic policies. Social Darwinism is based on the idea competition between individuals and groups ensure that superior people will prevail over inferior people. This kind of competition is viewed as necessary for human progress.
- Species: a group of closely related organisms capable of breeding together. Species is the basic category in the system of scientific classification developed by the Swedish naturalist Carl Linnaeus (1707-1778).
- Theory: a detailed explanation of some aspect of the natural world, based on a body of facts that have been confirmed through extensive observation and experiment.
- Transmutation: the theory of gradual organic change first developed by Jean-Baptiste Lamarck and Erasmus Darwin (1731-1802) in the early 1800s. The term evolution, which describes the same principle, was not in common use until late 1800s.
- Uniformitarianism: the theory that changes in the Earth's history took place gradually and were caused by the same slow-acting natural forces that can be observed in the present.

Index

Albert Einstein, 5, 6, 10, 11, 32, 70, 76, 80, 81, 82, 83, 84, 87, 88, 89, 90, 94, 97, 101, 102, 107, 114, 115, 118, 119, 121, 122, 123, 128, 130, 131, 132
Anaximander, *xi*
Archeology, *xi, xv*, 12, 29, 114
Bishop, *xv*, 81
Charles Darwin, ix, x, xi, xii, xiii, xiv, xv, xvi, xvii, 2, 3, 5, 7, 10, 11, 12, 16, 20, 21, 22, 23, 25, 26, 27, 28, 29, 30, 31, 33, 41, 42, 47, 48, 49, 50, 51, 54, 55, 56, 60, 61, 63, 64, 65, 66, 67, 68, 69, 70, 71, 72, 73, 77, 78, 79, 88, 92, 93, 123, 124, 135
Christianity, *xii*, 18, 82
Churches, *xii, xiii, xv*, 3, 42, 43, 44, 45, 46, 60, 61, 62, 63, 64, 65, 67, 71, 72, 126
Colonization, 12, 15, 18
Empedocles, *xi*, 112
Erasmus Darwin, *x, xii*, 137
Eugenics, *xvii*, 19, 20, 21, 22, 59, 135
Genetics, *xi*, 12, 26, 30, 55, 109
Greeks, *x, xi, xii*, 63, 81, 101
Gregor Mendel, 30

John Rigden, 122, 123
Paleontology, *xi*, *xv*, 11, 12, 24, 29, 46, 71, 114, 136
Physiology, 102, 110
Oxford, *xiv*, *xv*, 24
Richard Owen, *xiv*, *xv*
Rick Potts, *vii*, 52
Roman Empire, *x*, *xi*, *xiii*, 42, 62
Vatican, *xii*, 46, 60, 61, 63, 64, 65, 71

Footnotes

John S. Rigden, *Miracle*, (2005).

Charles Darwin, *On the Origin of Species by Means of Natural Selection, or the Preservation of a Favored Races In the Struggle for Life* (1859), p. 162.

Ibid, p. 158

Unlocking the Mystery of Life, (Illustration Media, 2002).

Michael Behe, *Darwin's Black Box*,(1996).

Michael Dentor, *Evolution: A Theory in Crisis* (1986), 250.

Charles Darwin, *On the Origin of Species by Means of Natural Selection, or the Preservation of Favored Races in the Struggle for Life*, (1859), 155.

Christoph Schonborn, *Finding Design in Nature*, (New York Times, July 7, 2005).

William A. Dembski, *Evolution as Alchemy*, (June 23, 2006). www.designinference.com/2006.

References

Born, Max, *Einstein's Theory of Relativity*. New York: Dover, 1965.
Calder, N., *Einstein's Universe*, New York: Penguin, 1988.
Clark, Ronald, *Einstein: Life and Times*, New York: Avon, 1984.
Dalai Lama, *Science and Spirituality*
Dyer, Wayne, W. *The Shift*, Hay House: USA. www.hayhouse.co.uk [7, 8, 9, 10 11, 12, 13, 14, 15, 14, 15].
Eldredge, Niles, *Darwin: Discovering The Tree of Life*, New York: W.W. Norton & Company, Inc.
Fleisher, Paul, *Matter and Energy*, Minneapolis: Lerner Publishing Company, 2002.
Fleisher, Paul, *Relativity and Quantum Mechanics*, Minneapolis: Lerner Publishing Company, 2002
Http://humanorigins.si.edu/evidence.
Morris, Simon, C., *Life's Solution*, Cambridge: Cambridge University Press.
Nardo, Don- *The Origin of Species: Darwin's Theory of Evolution*, San Diego: Lucent Books.
Quammen, David, *The Reluctant Mr. Darwin (The Making of His Theory of Evolution)*, New York: W. W. Norton.

www.ingramcontent.com/pod-product-compliance
Lightning Source LLC
Chambersburg PA
CBHW061509180526
45171CB00001B/100